RECUEIL

DE PLANCHES,

SUR

LES SCIENCES,

LES ARTS LIBÉRAUX,

ET

LES ARTS MÉCHANIQUES,

AVEC LEUR EXPLICATION.

SECONDE LIVRAISON, EN DEUX PARTIES.

PREMIERE PARTIE. 233 *Planches.*

A PARIS,

Chez
{
BRIASSON, *rue Saint Jacques, à la Science.*
DAVID, *rue & vis-à-vis la Grille des Mathurins.*
LE BRETON, *Imprimeur ordinaire du Roy, rue de la Harpe.*
DURAND, *rue du Foin, vis-à-vis la petite Porte des Mathurins.*

M. DCC. LXIII.

AVEC APPROBATION ET PRIVILEGE DU ROY.

A Diderot Pictorial Encyclopedia of
TRADES AND INDUSTRY

Manufacturing and the Technical Arts in Plates Selected from
"L'Encyclopédie, ou Dictionnaire Raisonné des Sciences,
des Arts et des Métiers" of

Denis Diderot

Edited with an Introduction and Notes by
Charles C. Gillispie

In Two Volumes
Volume One

DOVER PUBLICATIONS, INC.
New York

Published in Canada by General Publishing Company, Ltd., 30 Lesmill Road, Don Mills, Toronto, Ontario.
Published in the United Kingdom by Constable and Company, Ltd., 3 The Lanchesters, 162-164 Fulham Palace Road, London W6 9ER.

This Dover edition, first published in 1993, is an unabridged and unaltered softcover republication of the work originally published by Dover Publications, Inc., in 1959. The plates were selected and the introduction and notes were written by Charles Coulston Gillispie, and the index was compiled by the editorial staff of Dover Publications. Most of the plates have been reproduced facsimile size.

DOVER *Pictorial Archive* SERIES

Manufactured in the United States of America
Dover Publications, Inc., 31 East 2nd Street, Mineola, N.Y. 11501

Library of Congress Cataloging-in-Publication Data

Encyclopédie. English. Selections
   A Diderot pictorial encyclopedia of trades and industry : manufacturing and the technical arts in plates selected from "L'Encyclopédie, ou dictionnaire raisonné des sciences, des arts et des métiers" of Denis Diderot / edited with an introduction and notes by Charles C. Gillispie.
      p.   cm. — (Dover pictorial archive series)
   Reprint.
   Includes bibliographical references and index.
   ISBN 0-486-27428-4 (pbk. : v. 1). — ISBN 0-486-27429-2 (pbk. : v. 2)
   1. Industrial arts—Encyclopedias. 2. Industrial arts—History—Pictorial works. I. Diderot, Denis, 1713-1784. II. Gillispie, Charles Coulston. III. Title. IV. Series.
T9.E472513 1993
670—dc20   92-31820
                                                                                  CIP

Foreword

As will appear in the Introduction, Diderot's relations with his publisher, Le Breton, were chequered. Not only did Le Breton reap the profit and leave Diderot to subsist, as he wrote Voltaire, "on laurel," but he protected his investment in the work by violating his editor's confidence. I, too, have engaged myself to work upon *L'Encyclopédie* at the instance of a publisher. In justice to his initiative and imagination, I should like it to be clear that the credit attaching to the idea of reproducing these plates is his. His also are the decisions which are reflected in the technical quality of the reproductions. These I have not seen, but as we go to press I am confident that my experience will prove different from Diderot's in every important respect.

Nevertheless, we have had our little disagreement, the publisher and I, not I hasten to say over anything substantive, but over the trivial question of the title. The title I do regret, as unhappy in its phrasing and unhistorical in its implication. For Diderot never meant these pictures as an "Encyclopedia" in themselves, but only as illustrations to the great *Encyclopédie,* that distinctive achievement of the Enlightenment. The publisher feared, however, that my preference for a simple descriptive appellation would seriously damage his commercial chances. Since our arrangements do not give me an interest in the sale of the work, and since it is his book as well as mine, I feel that to penalize the publisher without whom it would not exist would put too high a price on my scruples about the name. Accordingly, I have agreed to explain that the title represents his conception of the book, and the subtitle mine.

I am deeply indebted to my wife, for her assistance in editing and for all the secretarial work. The selection reflects her taste. I dedicate my part in this book to her, and she joins me in apologies for any errors which, despite all our care, must certainly remain in a work covering so wide a range of technicalities.

September, 1958   CCG
                                                               Princeton, New Jersey

Second Printing

"I am indebted to Mr. Henry J. Kauffman for correcting a mistaken description which accompanied Plate 60 in the first printing, and to Professor Peter Gay for pointing out errors of fact and typography in the Introduction."

CCG
November, 1959

Table of Contents

Volume I

Introduction xi

Plates

Agriculture & Rural Arts 1-56

Art of War 57-81

Iron Foundry & Forge 82-121

Extractive Industries 122-164

Metal Working 165-208

Volume II

Glass 209-275

Masonry & Carpentry 276-303

Textiles 304-358

Paper & Printing 359-386

Leather 387-400

Gold, Silver & Jewelry 401-428

Fashion 429-446

Miscellaneous Trades 447-485

Index of Persons and Places

Index of Subjects Illustrated in Plates

*The purpose of an Encyclopedia is to assemble the knowledge scattered over the surface of the Earth; to explain its general plan to the men with whom we live and to transmit it to the men who come after us; in order that the labors of centuries past may not be in vain during the centuries to come; that our descendants, by becoming better instructed, may as a consequence be more virtuous and happier, and that we may not die without having deserved well of the human race.**

Diderot in the article "Encyclopédie."

Let us at last give the artisans their due. The liberal arts have adequately sung their own praises; they must now use their remaining voice to celebrate the mechanical arts. It is for the liberal arts to lift the mechanical arts from the contempt in which prejudice has for so long held them, and it is for the patronage of kings to draw them from the poverty in which they still languish. Artisans have believed themselves contemptible because people have looked down on them; let us teach them to have a better opinion of themselves; that is the only way to obtain more nearly perfect results from them. We need a man to rise up in the academies and go down to the workshops and gather material about the arts to be set out in a book which will persuade artisans to read, philosophers to think on useful lines, and the great to make at least some worthwhile use of their authority and their wealth.

Diderot in the article "Art."

* Reprinted from *The Censoring of Diderot's Encyclopedia* by Douglas H. Gordon and Norman L. Torrey through the kind permission of Columbia University Press.

Introduction

The serried backs of numerous Encyclopedias stretch like a yawn across yards of shelf in the reference rooms of libraries, and seldom stimulate those who exhume facts to think of information as alive. Its soporific quality does not prepare the student to understand the alarums and excursions which attended publication in Paris of the first great compilation to range knowledge along the alphabet, the *Encyclopédie, ou Dictionnaire Raisonné des Sciences, des Arts, et des Métiers*—Analytical Dictionary of the Sciences, Arts, and Trades—by a "Society of Men of Letters."[1] This was no innocuous work of reference. In the words of its moving spirit and chief editor, Denis Diderot, a good dictionary ought to have "the character of changing the general way of thinking."†[2] The first volume came off the press in 1751. Two years earlier Diderot had had to pass an uncomfortable and alarming hundred days as a crown prisoner in the fortress of Vincennes. There he was able to work on arranging the plates from which we reproduce a selection. He had been incarcerated, it appears, rather as a troublesome scribbler, author of the deistic *Philosophic Thoughts* and other questionable writings, than as editor of the projected *Encyclopedia*, but angry rumors about the threat posed by that undertaking to the "general way of thinking" were circulating ominously among policemen, princes, and priests.

"Up till now," wrote the Bishop of Montauban, a powerful priest, à propos of the publication of Volume II in 1752, "Hell has vomited its venom, so to speak, drop by drop. Today there are torrents of errors and impieties which tend towards nothing less than the submerging of Faith, Religion, Virtues, the Church Subordination, the Laws, and Reason."† What provoked his anger was the article "Certitude" by the Abbé de Prades, a freethinking churchman, who seated his subject in the

† Reprinted from *Diderot: The Testing Years: 1713-1759* by A. M. Wilson through the kind permission of Oxford University Press.

reasonings of John Locke instead of the mystifications of Revelation, and who—it scandalously developed—had just been granted a doctorate for a thesis arguing this view by the faculty of the Sorbonne, professorial guardians of orthodoxy asleep at their post. The offending abbé prudently departed for Prussia. The Council of State forbade continuation of the *Encyclopedia,* only to be overborne by the forces of liberal opinion in Paris and mollified by promises of self-censorship. In the provinces blasphemy was less clever and more dangerous. In 1766 the 19-year old Chevalier de La Barre, a foolish gay blade, was arrested in Abbéville for having desecrated a wayside shrine. He was said to have scoffed at the dogma of the Virgin Birth. His judges of the local *Parlement* decreed that his tongue should be removed, and his right hand cut off, after which he was to be beheaded. His body was to be burnt. The latter two parts of the sentence were actually carried out in the great square of Abbéville. It was the opinion of the court that his mind had been seduced by the influence of the Encyclopedists.

The *Encyclopedia* appeared before its readers in the complementary guises of ideology and technology. I have drawn upon the latter aspect for this book. Its purpose is to reproduce many of the magnificent engravings which illustrate how things were made and how people got their livings. But of the two aspects, the ideology is the better known. For it is the ideology of progress and liberalism, conceived in the 18th century flirtation of the French Enlightenment with the English Constitution, which burst into life in the French Revolution and matured into the system of verities of modern democratic society. What with the dismay and disarray into which the disasters of the last decades have plunged France and England, it may be only in America that these beliefs, even when honored in the breach, continue to be associated with the idea of progress. Tolerance, education, representative government, individualism, equality before the law, opportunity, material betterment—they make a familiar list. But these good things could not be espoused in 18th century France except in the negative form of criticism of a society which was quite otherwise: monarchical, hierarchical, and priest-ridden; where dignity and recognition were denied the ablest elements of the population who did the work and paid the taxes; where the road to political power led, not through statesmanship, but through the bed of a king reduced by implacable boredom to the last stages of idleness; and where intellectuals or "philosophes" (by their own account men of wisdom, virtue, and wit, for the latter two qualities do not exclude each other in the French scheme of things) could make no impression on the world except by writing books and contriving all sorts of shifts to get them around the censor.

Handling censorship was as essential to literary communication as handling syntax. Nor must the gaiety of expression in a Diderot or a Voltaire lead his reader to suppose that it was a sport unaccompanied by humiliation and some little danger. To judge how far one dared go was a nice thing. Miscalculation might land a writer in prison and his publisher in bankruptcy. Liberal ideology, therefore, had to be clothed in innuendo, irony, and indirection. So, for example, the article on salt is not confined to the properties of sodium chloride. It deals also with the injustice to the poor of levying taxes on the necessities of life. So, too, the article "Political Authority" opens: "No man has received from nature the right to rule others. Liberty is a gift from heaven, and every individual of the race has the right to enjoy it as soon as he attains to the enjoyment of his reason."† In another vein, the word "Fornication" is introduced as a term in theology, and the verbs, "Adore, Honor, and Revere" are distinguished in religious and civil usage: "In religion one adores God, honors the saints, and reveres relics and images; in civil life one adores a mistress, honors the worthy man, and reveres those who are illustrious and distinguished." The barbed cross-reference was an obvious device in the tactics of evading control. The reader is sent, for example, from an article on Belbuch and Zeombuch, gods of the Vandals, to one on Immortality and Soul. Tongue-in-cheek articles on religious and political practices recount absurdities in endless detail, to conclude with smug expressions of conformity so unctuous as to be utterly unconvincing. Not to multiply illustrations, however, there has recently been published for those who read French a brief and very useful selection of the pieces which epitomize the political, social, and religious ideas of the Encyclopedists.[3]

      But to make an absolute distinction between those ideas and the technical contents of the *Encyclopedia* is to miss the point. For Diderot's brilliant and original conception was to make the technology carry the ideology. This was his master stroke. Our own age sees science as power, but because relatively few people take the trouble to understand very much about it, we are less impressed with its purely intellectual triumph than was the 18th century. Over its sense of the world towered the figure of Isaac Newton, who had united knowledge of heaven and earth in a mathematical physics, and who symbolized the achievement of science, the new instrument of thought which had come into the world since the Renaissance with the unique capacity to be right about nature. Once understood, Newtonian science imposed assent with all the assurance of a geometrical theorem. In the *Encyclopedia*, therefore, a dictionary of science, it is not just Denis Diderot or Jean-Jacques Rousseau who seems to say that the institutions of the old Europe are absurd, immoral, and contrary to nature. It is science, and who is there to disagree except obscurantists and oppressors? In critical retrospect, there is, it is true, one little difficulty with this lesson. Diderot's conception of science was drawn rather from Francis Bacon than from Newton. But this made no difference to his success. Indeed, to invest the utilitarian idea of progress with the high authority of Newton was itself a feat, whether or not of legerdemain.

      No venture lacks forerunners, and the century-old philosophy of Francis Bacon worked as inspiration and example in the minds of the Encyclopedists. It inspired them by its summons to a new birth of knowledge enlisted in the service of humanity where learning would find food for healthy growth. Science was to serve mankind and not the pedant's pride or pocketbook. It set the example by seating the unity of learning in a classification of subjects according to their relations in nature. But Bacon's strategy of natural history was a straightforward attack on ignorance and futility, as it could afford to be in the England of the 17th century where conservatism was not obscurantist. For the devious tactics of camouflage, flanking movement, and Trojan horse, the Encyclopedists looked rather to Pierre Bayle, author of a *Historical and Critical Dictionary,* whose vein was different.[4] The work is at once a dictionary of biography and a set of historical attacks upon religious superstition and intolerance. Thus, the article on Pyrrho presents a sympathetic picture of that philosopher who "found in all things reasons to affirm and to deny; and therefore he suspended his assent after he had well examined the arguments pro and con . . . Hence it is

†from *Diderot* by A. M. Wilson.

that he sought truth as long as he lived." The article on David, on the other hand, brings out the trickery, cruelty, and lust of that ancient king of the Jews and suggests that in any but a great figure in religious tradition, his actions would be those of a scoundrel. Depending on the point of view, Bayle's *Dictionary* is a work of sacrilege or emancipation: either way it opened the Enlightenment in a spirit of critical skepticism.

Bacon and Bayle were distant progenitors. The entrepreneur was a publisher, André François Le Breton, a good bourgeois interested not in philosophical matters but in the opportunity to sell a compilation of technical information. Several such works had been published in England, the most successful being Ephraim Chambers' two-volume *Cyclopaedia, or Universal Dictionary of the Arts and Sciences* (1728). Originally, Le Breton thought only to publish a translation and enlargement of this work. He can have had no conception of the imbroglios into which his editors would lead him, nor of the lengths to which Diderot's imagination would go. His first translator, an Englishman called John Mills, had a command of the French language adequate for an attempted swindle but not for rendering Chambers' text. Their association dissolved in fisticuffs and canings in 1745 after less than a year's quarreling. We do not know exactly when or how Diderot was brought in, but it may have been at the instance of the Abbé Jean-Paul de Gua de Malves, a mathematician eccentric to the point of insanity. He, according to some accounts, had a fugitive connection with the enterprise as Mills' successor. In any case, Le Breton already saw the need for further capital and took three other publishers into partnership. It was well that he did so. The first volume did not go on sale until 1751. When Diderot got his head, the translation of Chambers' colorless and useful little work enlisted the collaboration of the foremost men of French letters, took twenty-five years, and ran ultimately to seventeen volumes of text and eleven of plates, followed by four supplementary volumes of text and another of plates. These additions, however, were edited by other hands than Diderot's, and published by the house of Panckoucke.

The Diderot who expanded a translating commission into the crowning achievement of the Enlightenment was thirty-seven years of age and a nearly unknown writer in 1750, when he published his Prospectus inviting subscriptions. By birth he was a provincial, a country boy born in the ancient and very Catholic city of Langres in Champagne, where his father, Didier Diderot, was a cutler whose scalpels had earned the confidence of surgeons. His mother's family were tanners. On both sides, therefore, Diderot was descended from people whose standing in the world was at the meeting place between artisan and bourgeois, a position comparable, perhaps, to that of the small businessman in contemporary America. This was precisely the level occupied by proprietors of the average enterprise pictured in the *Encyclopedia*. Diderot himself wrote the article "Cutlery," which is illustrated in Plate 179 (though not from his own father's shop). A rather elaborate tannery appears in Plate 389ff. But though he could admire, and sometimes sentimentalize, the artisan's honest toil, Diderot was not the man to share it. An attempt to work for his father ended in a return to his studies after a few days.

An omnivorous and insatiable student, Diderot went to school to the Jesuits, the schoolmasters of Europe, and even toyed with the thought of becoming a priest himself. This was not utterly improbable, for the church was one avenue to a lettered career and wisely left it to the conscience of the priest what measure of devotion to expend. Many clerics performed the motions required of them in a spirit which, like that of a modern scholar adhering to the pieties formulated by some state legislature, partook rather of the practical than of the heroic or the hypocritical. But it was no doubt fortunate for both parties that instead of taking orders, Diderot ran away to Paris. There his father, with real generosity, supported further studies, probably at the famous Collège Louis-le-Grand, under the impression that the boy was preparing himself for a legal career. Only when it became apparent that Diderot was as averse to the law as to the psalter or the grinding wheel, did his father withdraw support, after which Diderot disappeared into the limbo of Bohemia. There he spent his youth, not unhappily it would seem, emerging into the view of his biographers only in flashes of reminiscence, and living no one knows how. One commission is said to have been ghost-writing six sermons for an inarticulate missionary about to depart for the Portuguese colonies. He was in rags much of the time and inhabited a garret, from which he later remembered seducing a hardworking neighbor's wife simply by staring wordlessly but meltingly across the alley separating their lodgings, he shirtless and appealing in his window, she curious and mettlesome in hers. Out of this Beggar's Opera world of attics, doubtful cafés, omnivorous reading, and catch-penny writing, Diderot emerged a "philosophe," and a boon companion of d'Holbach, Grimm, and above all Rousseau, with whom his ties were closer than with any other until 1757, when Rousseau's growing morbidity, exacerbated by Diderot's untimely frivolity, brought a sad and angry parting. It was of this brilliant and unpredictable circle that the Marquis de Castries, a future Marshal of France, remarked that he could not understand it—these fellows kept up no estate; they inhabited lofts; and yet everyone talked of nothing but their doings.

Diderot was interested in the relation of everything to humanity. Among his many enthusiasms—the theatre, the fine arts, music, the classics—he reserved perhaps the keenest for medical science and lore. He had learned the language of Bacon, Locke, and Newton. It was, therefore, not simply as a literary hack that he undertook the translation into French of a three-volume *Medicinal Dictionary*, compiled and published in London by Robert James. It seems probable that this, in turn, is what recommended him to Le Breton. At the same time, Diderot was becoming known for other writings, not only known but—witness the Vincennes interlude—notorious. Only following his death in 1784, however, were the manuscripts of several purely literary masterpieces published. They had circulated through various devious channels, one of which led incongruously through the bedizened court of Catherine the Great of Russia, where Diderot went for a time as a pet philosopher in his later years.

Diderot was not primarily a creative writer or a systematic thinker. He was, in the largest and most favorable sense, a publicist, one who believed in his wares, a propagandist who worked for love of the game. All his writings are illuminated by quick perception and a highly personal response to issues. Thoroughly, even aggressively masculine in character and behaviour, his intellectual temper was almost feminine. His thought, therefore, escapes definition. One modern critic compares it to the parry and thrust of a fencing match. But his great quality was a sensibility keen but not narrow, highly polished, first by the Jesuits, then by self-education, until it was like a many faceted jewel catching and sparkling out to the reading public all the main themes of the 18th century Enlightenment.

Like the Enlightenment itself, Diderot began with Locke. *The Letter on the Blind* develops the sensationalist psychology according to which man is what he makes of his experience. *The Nun* attacks the practise of disposing of unwanted daughters to the convent, where they were doomed to lives of inane sterility. *Philosophical Thoughts* advances the deist case that a religion of reason and humanity is the best. The *Bijoux Indiscrets* (Naughty Jewels) is a work of mild obscenity, a dirty book written for money or for fun, perhaps for both. Though he published minor mathematical papers, he parted company with Voltaire and the rationalists in his dismay at the irrelevance to the human condition of analytical and descriptive Newtonianism, that is to say of classical mathematical physics. His *Thoughts on the Interpretation of Nature* followed by the *Dream of d'Alembert* were among the first writings to protect a sentient nature from the doom of physical science (which kills the quality of what it measures and describes) by cloaking reality in the biologist's veil of process and organism, under which nature can be thought continuous with personality. This romantic defense of an intimate nature, sympathetic to man, was to prove a lost but not a forlorn cause in the next century, though since Darwin it only flutters ineffectually in the breasts of those who love nature but resent science.

Conversation is a great art in France. It has little in common with the leaden exchanges which thump about in drawing rooms where English is spoken, and the *Dream of d'Alembert* is an imaginary exchange between d'Alembert, Mlle. de l'Espinasse, his mistress and a great hostess, and Dr. Bordeu, a famous physician and man of wit, wisdom, and prescience. Amongst them they sketch out an astonishing vision of a sentient universe in evolution, its matter pulsing with consciousness like a cosmic polyp, with which to replace the deadly world of physics containing only bodies in motion. This is a fanciful anticipation of the Lamarckian theory of evolution (though it has nothing important in common with the Darwinian theory). In a letter to his mistress, Sophie Volland, Diderot explained why he chose to put such startling ideas in the mouth of a dreaming mathematician: "Wisdom must often be given the air of madness. . . . I would rather people say, 'But that's not as crazy as you might think,' than tell them, 'Listen, what I am going to say is very wise.' "[5] Except Plato and possibly Galileo, no one has excelled Diderot in the dialogue. He is at the top of his disturbing form in the satire *Rameau's Nephew*. The title character, a ne'er-do-well relative of the musician, is half genius, half *clochard*. He fascinates the interlocutor, Diderot himself, by the outrageous degradation of his squandered talent. But is the fault in him or in the society which casts the artist as clown and parasite? (Diderot, who has been variously claimed as a forerunner of Goethe, Marx, Darwin, Freud, and Joyce, might equally well be said to have anticipated the tragic buffoonery of Dylan Thomas). And we recognize our secret springs of action in the inadmissible views which the ragged Rameau expresses on life and morals. Diderot's indignation is eminently unconvincing and skillfully calculated to reveal glimpses of the hypocrisy behind the mask of respectability. This is Diderot's literary masterpiece. It is a complicated, somewhat diabolistic book.

Jacques the Fatalist anticipates more straightforwardly the novel of realism and naturalism, and—not to catalogue Diderot's entire works—in *Supplement to the Voyage of Bougainville* a Tahitian chieftain offers his daughters to the young priest accompanying the expedition. This frank and kindly hospitality converts that functionary, beneath whose cassock beats a full-blooded heart of gold, to the naive decency of free love in an unspoiled society. So it was that each of Diderot's writings served as a vehicle in praise of some aspect of the natural man. He cannot be assigned to either the rationalist or the romantic moods. He belonged to both and to neither. Intelligent, amorous, sensuous, cultivated, critical, sentimental, humane, hard-working, humorous and pagan—he was in his own eyes above all else virtuous, in that his life and writings were affirmations of the innocence of nature—of nature in general, of his own nature in particular, and of the opportunities offered by their congruence.

But despite Diderot's virtuosity (or because of it), his achievement of the *Encyclopedia* is what brings him forward from among the crowd of interesting writers to the front rank of French letters. He had collaborators, of course, although all the volumes were edited in his own lodgings without secretaries or office paraphernalia. The foremost was the eminent mathematician, Jean-le-Rond d'Alembert, whose views in real life were not at all those of the *Dream*. D'Alembert's reputation was already established, as Diderot's was not. His ancestry was also more distinguished. Both parents belonged to aristocratic families. The only difficulty was that they were not married. Nor could this be remedied, for though his father, the Chevalier des Touches was a dashing cavalry officer, his mother, Mlle. de Tencins, was a nun. She had abandoned the baby on the steps of the church of St. Jean-le-Rond, to which words he added the surname by which he is known. His father saw to his upbringing and education. As he outlived what has been called that "athleticism of the intellect" essential to creative mathematics, his mind mellowed and softened as the minds of mathematicians will, and he turned his attention to philosophy and the problems of humanity in society, which meant the same thing in the French Enlightenment.

D'Alembert, then, did not confine himself to supervision of the mathematical and scientific articles. On the contrary, it was he who wrote the Introductory Essay, which summarized the encyclopedic ideal of a progress to be achieved by unifying knowledge in the service of mankind. This piece created a great impression and remains a cardinal document of the Enlightenment. But though much more than a simple contributor, d'Alembert turned out something less than a co-editor. More ambitious than Diderot for official status, he tended to oscillate between indiscretion and alarm. His article on Geneva in Volume VII provoked an acute crisis. In implied comparison to France, that Protestant city-state was represented as a model of civic virtue, spotted only by the universal vice of intolerance which there took the form, incomprehensible to a Parisian, of a ban on the theater.

Not only did the article give offense to the authorities as a critical Utopia, but it embarrassed them as an unwarranted interference in the sumptuary laws of a neighboring republic. Moreover, it compromised the Calvinist ministers of Geneva by suggesting, in the guise of a compliment to their good sense, that they were at heart Unitarians. This was an unlucky article from every point of view. It strained to the breaking point the fragile connivance by which the chief censor (of all people) had been protecting the *Encyclopedia* from the scurrilous and rabid campaign mounted by clerics and reactionaries. For Lamoignon de Malesherbes, who held that office, was a young and liberal nobleman, a believer in regulating the press so as to assure its freedom so far as possible and politic. He was one of the heroes of the story.

But "Geneva" made the situation impossible, coming at a time when the authorities were alarmed by other subversive books and portents. While Volume VIII was still in press, the Council of State, the highest administrative body in the land, took the matter out of Malesherbes' hands and decreed suppression and seizure of the *Encyclopedia,* together with other dangerous writings. Malesherbes managed to warn Diderot that a search of his notes and papers was impending, in time for him to hide them until the storm blew over, as ultimately it did. But all these menaces and rumors proved too much for d'Alembert, who henceforth dissociated himself from the *Encyclopedia,* putting it about that the times were not ripe. He was right, of course, but his attitude left it up to Diderot to ripen them, to honor the subscriptions, and to save the publishers from losing the large sums invested. It is consistent with Diderot's gaiety of mind that he should not have reproached d'Alembert too openly for desertion, nor resented the prestige which had earlier accrued to d'Alembert as author of the Introductory Essay and as the more philosophical collaborator.

It was Diderot, the scribbler, rather than d'Alembert, the scientist, who displayed fidelity to his engagements and admirable persistence. From the outset he surmounted the insurmountable. He assembled writers and assigned subjects. He arranged for the engraving of plates. It is one instance of the attention required by detail that these were not printed until many years after the articles, yet in only one case (passementerie) do the key-letters which refer the reader from text to plate fail to correspond. He blue-pencilled overlong articles and padded the overshort. He got contributions out of Rousseau, out of Montesquieu, out of d'Holbach, out of Saint-Lambert, out of Condorcet. He found a workhorse, the Chevalier de Jaucourt, who would write in the Encyclopedic idiom on anything, and not badly. He dealt with Voltaire's coquettishness. When the success of the early volumes transformed that great man's early disdain into a desire to participate, Diderot delicately handled this waspish prestige so that it augmented rather than diminished the Encyclopedia's. He kept an eye on the censor, resigned himself to weakening his effect, and withstood the rising clamor of clericals and conservatives.

Nor must it be supposed that Diderot and the liberals had a monopoly of wit. The damaging term "Cacouacs"—Quack-Quacks—was coined for the Encyclopedists by the right-wing journalist Moreau, whose kind passionately pursued them with charges of plagiarism and impiety and sought to envenom those prejudices of the ignorant to which intellectuals and rich men are, in their different ways, peculiarly vulnerable. Yet it was the publisher to whom he had been loyal who, with the best intentions, struck Diderot the lowest blow. Ten volumes covering the letters "H" to "Z" were still to appear after the "Geneva" crisis. Apprehensive about his investments, not trusting Diderot's judgment, Le Breton began himself to delete passages which he feared overbold, but only after Diderot had corrected and passed page-proof.[6]

Yet even when he discovered this treachery, this surreptitious mutilation of his own articles and those of many contributors, Diderot was determined to complete the *Encyclopedia,* and he did so, not on the best terms, but on the best terms he could get. The extent of Diderot's responsibility for the *Encyclopedia* is a question that has been much discussed among literary historians. For a time the tendency was to denigrate his efforts, to say that the original idea was not his, that the philosophy was Bacon's or Locke's, the science d'Alembert's, the psychology Condillac's, the wit Voltaire's, the feeling Rousseau's, that Diderot worked with scissors and paste rather than ideas of his own. Recent scholarship appreciates his merit at a higher and juster level. As a bicentennial commemoration sensibly points out, Diderot's contribution to the *Encyclopedia* was that he produced it.[7]

To the Anglo-Saxon who admires French culture but regards French politics with a certain dismay, it seems one of the misfortunes of that fearful game that every interest instantly perceives, with a Gallic clarity excluding compromise, the reality of some threat masked by the humane appearance of its opponents' projects. So it was in the 18th century. The enemies of the *Encyclopedia* were quite right to fear it. Knowledge was not innocuous, not when it was democratized, and that is the most important feature of the plates reproduced in this book: they represent democratization of systematic learning. The term may, perhaps, seem inappropriate, for "democracy" refers to government, not to scholarship, and in 18th century France the philosophic movement of opinion was liberal in tenets without being democratic in politics. But no one was, before the Revolution of 1789, and by planting in men's minds criticisms of the existing order and belief in the possibility, indeed the certainty, of a better one to be achieved simply by taking thought, the Encyclopedists prepared the Revolutionary mentality without foreseeing how it would change the government of the world.

It is in a sense deeper than the political, that Diderot's *Encyclopedia* was truly popular. From this point of view, it is the technology which is the fundamental aspect of the *Encyclopedia* and not the ideology. For the sardonic and skeptical influence of the *Encyclopedia* could never move beyond criticism of what was, to evocation of what should be. Nor could it ever touch any but the literate and educated, aristocrats flirting thrillingly with liberalism, intellectuals who already believed what it flattered them to be told cleverly. Wit is never really popular. The people sense its malice. It reaches them, if at all, only to make them uneasy and unsure of themselves. What was democratic in the *Encyclopedia,* therefore, was not the persiflage. It was the technology, the dignification of arts and trades. By taking craftsmanship seriously Diderot's *Encyclopedia* set values in motion which have indeed taught the artisan to think well of himself. For when they did not joke, the skeptical Encyclopedists preached, and the subject of their great long sermon was the dignity of labor.

In general the humanitarianism of the Enlightenment secularized and put to work assertions which had originated in Christianity. The contempt for manual labor which Diderot ascribed to the liberal arts was a Greek legacy revived in the Renaissance and instilled into the aristocracy by humanistic education in the classics. Nor was the idea new that philosophy is words while true knowledge is embodied in the skills of the artisan, that the blacksmith knows metals and the metallurgist only books. Both Roger Bacon in the 13th century and after him Francis Bacon in the 17th had urged philosophers to go to school to craftsmen.

It would have been easy to repeat this injunction. Instead, Diderot obeyed it. In doing so, however, he quietly and perhaps unwittingly switched roles of pupil and schoolmaster as between science and industry. One frequently reads of science as the fruitful element in western technology, as the progenitor of industrialization. So it has been, but not in the sense often implied. Basic theoretical science had very little to offer industry in the 18th century. The law of gravity, after all, could be admired but neither used nor evaded. It was descriptive science addressed to industry which first dignified and then transformed it by rationalizing and publicizing processes, and that was the importance to technology of the *Encyclopedia* as a dictionary of science. The different trades and industries were investigated and classified according to the principles revealed by the investigation. Thus, these hitherto obscure matters were brought within the unity of the human understanding. Henceforth, dyeing is not just a set of rules of thumb. It depends on a particular science, chemistry, one of the natural sciences which form part of philosophy, that branch of knowledge drawn from the reason. So, too, the manufacture of mirrors is an aspect of glassmaking, a mechanical art, one belonging to extractive industry, which is dependent on natural history, itself one division of history, that branch of knowledge drawn from the memory. Thus, all the arts can be seated in one of the three human faculties, reason, memory, or imagination, the latter ruling over the fine arts.

It is true that for convenience the *Encyclopedia* is arranged alphabetically rather than schematically. But the act of analysis exalted the trades. It raised them from lore to science. Some of the crabbed processes depicted in these volumes—the third beating of gold-leaf in the hide taken from the abdomen of certain cattle—could scarcely withstand the light of day. The *Encyclopedia* took a giant step toward replacing the Gothic instinct that techniques are trade secrets, mysteries to be concealed by the practitioner, with the concept of uniform industrial method to be adopted by all producers.[8] "To the tableau of the sciences," wrote Condorcet of this process of rationalizing publicity, "should be united that of the industrial arts which, leaning on science, have progressed more steadily and broken the chains in which routine had hitherto bound them."[9]

Science in the Enlightenment was the educator of industry, though not the source of its techniques. It needed education and, as happens in that situation, sometimes resented the process. Diderot's posture before the artisans exhibits a little of the ambivalence of those Russian intellectuals of a hundred years later, the "Narodniki," who in gusts of sentimental guilt went to the peasants, rude but richly human through centuries of suffering, to drink of virtue at the springs of folk wisdom, and who on arrival were greeted not by a great-hearted Russian people, but by churlish clodhoppers in a mood of canny suspicion none the more agreeable for being shrewd. Diderot's disillusionment was less severe, but expressions of impatience did escape him. It proved extremely difficult to find out about a great many trades. Language alone presented enormous difficulties. Each trade had its own barbarous jargon. For some a whole lexicon was needed. Worse, a great many artisans neither understood what they did nor wanted to understand. In their mulish way they preferred working by rote. "It is only an artisan knowing how to reason who can properly expound his work," exclaimed Diderot (a complaint which might seem to render circular the conception of the function of the *Encyclopedia*). While freely acknowledging numerous defects, Diderot challenges the inevitable detractor to do better: "He will learn by his own experience to thank us for the things done well and pardon us for those done ill. Especially will he learn, after having for some time gone from workshop to workshop with cash in his hand and after having paid dearly for the most preposterous misinformation, what sort of people craftsmen are, especially those at Paris, where the fear of taxes makes them perpetually suspicious, and where they look upon any person who interrogates them with any curiosity as an emissary of the tax farmers, or as a worker who wants to open shop."†

A picture speaks what words conceal, and although the text of the *Encyclopedia* carries long accounts of the principles and practices of crafts both major and minor, descriptions alone would have been arid, and would have fallen far short of the purpose. The plates to which most of the articles are keyed contain the essential record of 18th century technology. But it is precisely in relation to the origin of the plates that the nastiest accusations have been brought against Diderot's publishing ethics. Publication of the text, it will be remembered, was interrupted with volume VII (1757) by withdrawal of the license (in 1759). This was to be restored only on the understanding that the remaining volumes—ten as it turned out—would appear (in 1765) as a unit, in order to facilitate control of their contents. Meanwhile, there could be no objection to going ahead with the technical plates, which contained no sensitive matter, and publication of a *Collection of 1,000 Plates . . . on the Sciences, the Liberal Arts, and the Mechanical Arts* in four volumes (to grow to eleven) was announced to the subscribers in 1759, the first volume reaching them in 1762.

It was preceded by ugly charges of plagiarism, extending to the execution as well as the conception of the work. The French Royal Academy of Science (with the Royal Society of London one of the two most venerable and distinguished scientific bodies in the world) had been founded in 1666 under the aegis of Colbert, Louis XIV's great minister of state. Like most statesmen, Colbert took a thoroughly Baconian view of science, and in 1675 fixed upon the new Academy the responsibility of undertaking a scientific investigation and description of French industry. This work had been languishing in academic good intentions for eighty-six years. Its former director, the naturalist and metallurgist Réaumur, was a perfectionist who had not published a line. But he had commissioned a large number of plates. No sooner had Diderot's publisher announced their *Collection* than enemies came forward with injured complaints first put about by Réaumur (who died in 1757) that Diderot's agents had seduced Réaumur's engravers, that they had secured access to the proofs of the plates which were the property of the Academy, and that Diderot's illustrations were copies of those prepared for its *Description of the Arts and Trades*.

Georges Huard, who has recently investigated these charges, hitherto dismissed by admirers of the *Encyclopedia* as typical Jesuitical calumnies, concludes that there is much substance in them.[10] He goes further and points out that

†from *Diderot* by A. M. Wilson.

Diderot was not the first to carry science to the workshops—that members of the Academy were there before him. As to the latter argument, there can be no disagreement, though its edge is, perhaps, dulled by the reflection that Bacon was there before both, and that what finally brought the Academy to break the ice of publication was the threat of being forestalled. The first volume of the Academy's own series was rushed into print in 1761, a year before Diderot's first plates.

It is less easy to pass judgment on whether Diderot's unauthorized borrowing of the Academy's designs constituted plagiarism. Similar charges were laid against many of the articles. But the line between plagiarism and consulting authorities was not drawn very fine in the 18th century. One must admit that it would have been quite characteristic of Diderot's demi-monde of scribblers to wine, dine, and bribe Réaumur's faithless assistants into parting with proofs in order to have the designs re-engraved with slight alterations by Diderot's own stable of artists. Moreover, he must have planned on it from the outset, for the first seven volumes of text contain reference letters to the plates in question. Nevertheless, the conditions imposed on publication by the authorities turned it into a game like love or war where much was fair, if not all, and Diderot's most high-minded critics will agree that wherever the plates came from, he it was who got them between covers. At this distance, therefore, we need not, perhaps, pursue too closely the question of authorship. For we are not in any case concerned with these engravings as documents in the history of art (though they seem very fine and may merit study from that point of view—what other encyclopedia has ever been illustrated so well?) but rather as records of industry and life.

What of them, then, as records? It is immediately apparent that they are uneven. The glass industry, for example, is splendidly illustrated—one could take the text and plates, construct a plant with divisions for plate, smallware, and crown glass, and produce 18th century glass. Textiles, on the other hand, are not so well handled. Gobelins tapestries are handsomely covered, but the humbler operations of spinning and weaving are scattered throughout the volumes. The plates depicting them are repetitive, pesky, and unclear in the detail and structure of handloom and spinning devices. In the iron industry, smelting and forging are thoroughly covered, casting is depicted rather cryptically, and rolling and slitting are badly slighted. For all his pride of progress, Diderot's use of designs prepared many years earlier for the Academy dropped him well behind the times here and there. So, too, did his going to the artisans rather than the experts. On occasion, he even fell back on the 16th century, copying two plates from the 16th century metallurgist Agricola and (in the anatomy section not reproduced in the present volumes) a number from Vesalius *On the Fabric of the Human Body* of 1543. Elsewhere, as if to strike a balance, Diderot's parade of modernity was better than the last word, and he included drawings of machines which, like those of Leonardo da Vinci, existed only in the minds of their inventors.[11]

Nevertheless, it would be churlish to begrudge Diderot the expedients by which single-handed he put together uncounted articles and eleven volumes of technical drawings. With the Academy's *Description of the Arts and Trades,* from which he borrowed, and the *Encyclopédie méthodique,* beginning in 1782, which unimaginatively corrected many of Diderot's errors and extravagances, the illustrations of the *Encyclopedia* provide a unique record of a century of technology.

At their best, Diderot's plates have a sweep and stylishness which put them at the summit of the genre of technical illustrations. It is a genre which goes back in inspiration and method to the Renaissance, to Vesalius and to Leonardo da Vinci. In Leonardo's sense of form, mechanics and anatomy are expressions of a single marvelous vision of the world, which simply *sees* how things are. There is no more practical example of the importance of Renaissance naturalism than the illustrations of Vesalius's *On the Fabric of the Human Body*. (J. B. de C. M. Saunders and Charles D. O'Malley have recently reproduced these in a handsome edition, and the reader may be interested to compare it to the *Encyclopedia*).[12] Vesalius gives us the human body in two series, front and back views. First, the skin is flayed away to show the muscular structure. Then the student penetrates ever deeper through successive layers of muscle to skeletal structure and ultimately to the individual organs dismounted for his inspection. So it is with Diderot's technique of illustration, which is in inspiration and effect an anatomy of machines (a further instance, by the way, that it was descriptive rather than theoretical science which was applicable to industry). Typically, we are given first an over-all picture of an installation, then two sections at right angles, one lengthwise and the other crosswise. Thereafter, we penetrate by cutaway views to the essential assemblies, shown in place or in isolation, until finally we come down to the individual parts (or organs). To have shown all this in full would have surpassed the limits of the present selection and exhausted the interest of the modern reader in the detail of obsolete mechanisms. But the identification under each plate will lead the specialist to the last nut and bolt in the original.

The plates illustrate a state of industry which immediately preceded that rapid complex of developments known to text books as the industrial revolution. This capital event has recently been the subject of a considerable scholarly controversy, turning on two points. First, did it occur at all? Second, whether evolutionary or revolutionary, was industrialization good for mankind? These are questions which have been agitated very deeply by scholars drawing on vast funds of economic data, social conscience, and political disagreement.[13] It would be presumptuous to think of settling them on the evidence of a picture book. Nevertheless, certain impressions do suggest themselves which it may be appropriate to set out, and first of all about the question of revolution. This is not just a matter of rate. No one doubts the fast pace of industrial development after 1760, or that industry became the forward sector of the British economy by the 1830's. France, Germany, the United States, Japan and Russia followed in turn, and China and India will be next. But the problem of the point at which rapidity of evolution becomes revolution is metaphysical.

The proponents of revolution hold rather that industrial changes were in kind and not simply in degree, and they point to three factors: the substitution of iron for wood and stone as the fundamental structural material; the shift to factory production by an extension to all industry of externally powered machinery of the type introduced into the English textile trades by Richard Arkwright; and James Watt's invention of the separate condenser which turned the steam engine into a practicable prime mover. (In recent years engineers, who always tend to think they were born yesterday, have also been fascinated to find in Watt's governor the "feedback" principle, and hence automation). Reflection on these plates suggests that of these three factors, the steam engine was the truly revolutionary element. There is no more beautiful example of slow and steady evolution in technique than that offered by the history of the iron industry. The shift from charcoal to coke was an essential change of fuel, but the modern steel plant is the 18th century foundry writ large, as that in turn was the forge diversified (Pl. 86ff.). As to the factory system, Arkwright's cotton mills embody no principles beyond those to be observed in the paper factory of l'Anglée (Pl. 359ff.) or the Piedmont mill which had been throwing silk in Italy for centuries (Pl. 316).

Nor is the industrial dream itself a new vision. Eighteenth century experts were as aware as any latter-day technologists of the merits of rationalization, division of labor, standardization of parts, and displacement of uncertain labor by certain machines. But what they never foresaw was power, the exponential increments of power, in the steam engine, then in electricity, then in the internal combustion engine, next in the atom. Power was new, going infinitely beyond what could be pullied, levered, geared and screwed out of the forces of man and animal, wind and falling water. Power was the truly revolutionary instrument lying to hand, which has so transformed the world that hundreds of millions of men otherwise doomed by material insufficiency have lived.

The Encyclopedists' dream of a progressive industry illuminated by rational technique has been abundantly surpassed. What would be their astonishment, then, to read certain of their successors among modern men of letters, keening over culture and craftsmanship, who complain of the success of the dream in clichés like vulgarization, Americanization, subordination of man to the machine, or whatnot? "How bizarre is the working of the human mind!" (as Diderot once wrote), "Is it a question of discoveries? The mind distrusts its powers. It stumbles in self-created difficulties. Things seem impossible to find. But are they found?—It no longer sees why it had to look for them, and falls into self-pity."

But it is not just a question of men of letters. The social conscience has bitten deeply into history, and the belief is very widespread that industrialization came initially as a catastrophe. In part this reflects the influence of Karl Marx. His picture of independent artisans turned into wage-slaves has possessed generations of students who got it from a text book but do not know where the author got it (even when he does himself). Historically, however, this picture of laborers herded impersonally into mushrooming factories is related to the very concept of the industrial revolution itself, an idea only about eighty years old which was invented by the elder Arnold Toynbee (uncle of the historian) and a school of Christian socialists who imparted to the British Labor Party much of its idealism. Just as recent research in economic history has tended to spread industrialization much more gradually along the decades or even centuries, until the notion of revolution is attenuated to vanishing,[14] so the idea that the standard of living of the working class suffered from mechanization has been questioned, and in the present writer's opinion, overthrown. Not that work and life were not wretchedly hard for unskilled labor in the late 18th and early 19th centuries, but they always had been, and there is evidence that mechanization brought easement rather than impoverishment and subjection to the machine. Both Marxism and the more moderate socialist indictment of industrial capitalism have always turned on a rather halcyon view of craftsmanship and on an exaggeration of the independence of the artisan from the capitalist. And it is just here that the plates of the *Encyclopedia* bear on the question. For they give little encouragement to sentimentalize the work they display.

There was, certainly, beautiful handwork in the old Europe. It is exemplified in this book in crafts like enamelling, jewelry and cabinet-making, where something of loving care may be discerned in the attitude of the workers. (Though it must be remembered that the strictness of the regulations had much to do with the results, as it does once again in the contemporary French wine industry, where growers love their work but are held to standards). The plates themselves are a witness to the sort of work in which the artisan could be an artist. The words have the same root, and the French "artiste" may mean both. There may probably have been more of creativity then than there is now. But at best how slight a sector of life it filled! Most of what we see is hard, back-breaking, routine-bound labor for long hours with clumsy tools and little to show for it. Neither are we afforded many glimpses of Marx's independent artisan, enjoying the dignity of self-reliance through ownership of his tools. These workmen—most of them—whether in large foundry or small shop—look not less, but vastly more dependent on their employers than is the labor force of modern France, England, or America. The Encyclopedists foresaw the much deplored displacement of man by the machine with optimistic enthusiasm. They embraced the prospect as one of plenty, and as a liberation from the brutalizing drudgery to which tradition-bound routine condemned the peasant on the land and the laborer in the shop.

But that is by way of digression: it goes far beyond the purpose of the present publication to restore economic history to the optimistic mood of the *Encyclopedia* itself. It is offered, not as an analysis, but simply as a panorama of 18th century France at work. Even for this a word of warning will be wise. The scene, if not idealized, has certainly been tidied a bit by Diderot's artists, in accord, doubtless, with his enthusiasm for the arts and trades, and in deference to the sensibility of an 18th century public which preferred its portrait without warts. But this can be redressed with a little vicarious and therefore pleasurable discomfort, if the reader will exercise his other senses through the eye: if he will listen in the mind's ear to the cacophony of the forge, if in imagination he will smell the lime and offal of the tannery, if he will let his skin shrivel in the heat of the glass furnace. And occasionally the vast Hogarthian reality glares through the veil. It does so, for example, from the underside of the handicrafts in the basket-weaver's cellar (Pl. 459), and again in the shaft of the slate mine (Pl. 159). No machine sullies those picturesque scenes. But a few dips into squalor will suffice. Traveling in a far land, a little slumming goes a long way, and the main impression to be brought back from this trip through 18th century France is of the people who lived and worked there, not the scented aristocrats (though without them the picture would be incomplete, and we do have glimpses of dueling and powdering of hair), but the substantial people. They are strong, industrious, and competent, these manufacturers and shopkeepers. They know what they are doing and they do it well. Lawyers came arguing to the surface of politics in the great Revolution which loomed over this country. But these are the people who gave it body and who in every land built the world we all inhabit.

Diderot was acutely aware of the imperfections of his *Encyclopedia,* its mistakes, omissions, and infelicities of arrangement. He disarmed criticism by presenting it as a first attempt, to be superseded as soon as practicable by successors who would profit from his mistakes. It will, therefore, be gracious to take this, rather than Diderot's cavalier treatment of his own sources, as a license to make the omissions and rearrangements practised in the present selection. Unlike him, the present editor has taken pains to cite the location of each plate in the original volumes. Eight plates are reproduced from the supplementary volume published by Panckoucke, *Suite du Recueil de Planches sur les Sciences, les Arts Libéraux et les Arts Méchaniques, avec leur Explication,* which was not edited by Diderot. They complete the illustration of certain crafts. The arrangement, too, reflects a modern sense of continuity. In the *Encyclopedia,* certain industries—notably glass and iron—are presented *en bloc* in the order in which raw materials were worked up into finished product. The basic organization was alphabetical, and the operations of other trades, tanning, for example, and weaving, are scattered under the initial letters of subsidiary processes. As already indicated, and as is obvious from the scale of the original, the omissions have been extensive. The *Encyclopedia* includes some 2,900 plates, mainly on arts and crafts, but also on the sciences, engineering, the fine arts, sports, heraldry, and other subjects.

In choosing, the editor has been guided by the desire to achieve continuity and to preserve human interest. Hundreds of plates of the nature of blueprints have been omitted. In illustrating the smaller crafts, which sometimes took only two or three plates, often only one, the *Encyclopedia* gives a vignette of the shop in the upper half of the page, with an enlargement of the characteristic tools of the trade below. Only where these tools have a special interest, or where their detail is essential to understanding the trade, are they reproduced. Otherwise, the vignette alone is given. Perhaps it is a pity that the whole *Encyclopedia* cannot be simply reprinted. But the editor consoles himself for his mutilations by reflecting that it is available in libraries to readers who cannot have too much of a good thing, and he will justify human interest as a principle of selection in Diderot's own words: "Why not introduce man in our work, as he is placed in the Universe? Why not make of him a common center? . . . . Man is the sole and only limit whence one must start and back to whom everything must return, if one wishes to please, interest, touch, even in the most arid considerations and the driest details."†

†from *Diderot* by A. M. Wilson.

Footnotes

1. In 1951 the bicentennial of Volume I gave occasion for a number of scholarly commemorations. The Bibliothèque Nationale in Paris published the catalogue of its exposition under the title *Diderot et l'Encyclopédie* (Paris, 1951) to serve as a chronology and bibliography of the great venture. In addition, two learned journals devoted special issues to papers on the *Encyclopedia: "L'Encyclopédie* et le progrès des sciences et des techniques," *Revue d'histoire des sciences et de leurs applications* (1952); and *Annales de l'Université de Paris* (Numéro spécial, Octobre 1952), printing "Conférences faites à la Sorbonne à l'occasion du 2ᵉ centenaire de *l'Encyclopédie* du 3 mars au 23 avril 1952."

2. There has recently appeared an excellent biography covering the first half of Diderot's life: Arthur M. Wilson, *Diderot, the Testing Years, 1713-1759* (New York: Oxford University Press, 1957). Mr. Wilson's second volume will be eagerly awaited. The reader who turns to Mr. Wilson will perceive how much I have profited from his narrative. I have also taken the liberty of adopting some of his felicitous translations of certain passages from Diderot's writings which I quote in the Introduction.

3. J. Lough, *The Encyclopédie of Diderot and d'Alembert, Selected Articles* (Cambridge: At the University Press, 1954).

4. Two books of selections will exemplify the nature of the influence of Bacon and Bayle: Francis Bacon, *Essays, Advancement of Learning, New Atlantis, and Other Pieces,* ed. Richard F. Jones, (New York: Odyssey, 1937), and E. A. Beller and M. du P. Lee, Jr., *Selections from Bayle's Dictionary* (Princeton: Princeton University Press, 1952).

5. Diderot, *Oeuvres philosophiques,* ed. Paul Vernière, (Paris: Garnier, 1956), p. 251. This is the best edition of Diderot's philosophical writings. It accompanies a selection of his fictional pieces also published by Garnier, *Oeuvres romanesques,* ed. Henri Bénac (1951).

6. Douglas H. Gordon and Norman L. Torrey, *The Censoring of Diderot's Encyclopedia* (New York: Columbia University Press, 1947). This is the best account of the vicissitudes of publication. Mr. Gordon came into possession of a set of the *Encyclopedia* which included much of the proof corrected in Diderot's own hand. This has enabled the authors to confirm the fact of Le Breton's mutilations and to re-establish portions of the text.

7. Jean Thomas, "Le rôle de Diderot dans *l'Encyclopédie,*" *Annales de l'Université de Paris* (Numéro spécial, 1952), p. 25.

8. I have myself discussed this question of science and industry at greater length and with a certain amount of technical detail in two articles in Volume 48 of the journal *ISIS:* Charles C. Gillispie, "The Discovery of the Leblanc Process," and "The Natural History of Industry," (June and December, 1957), and I have pursued what seem to me some of the less happy implications of Diderot's Baconianism in a longer paper, "The *Encyclopédie* and the Jacobin Philosophy of Science: A Study in Ideas and Consequences," *Critical Problems in the History of Science,* ed. Marshall Clagett, (Madison: University of Wisconsin Press, 1959).

9. *Esquisse d'un tableau historique des progrès de l'esprit humain* (Paris, 1795), p. 296.

10. "Les planches de *l'Encyclopédie* et celles de la *Description des arts et métiers* de l'Académie des Sciences," *L'Encyclopédie et le progrès des sciences et des techniques* (note 1, above), pp. 35-46.

11. Bertrand Gille discusses the merits of the *Encyclopedia* and points out its shortcomings with somewhat more asperity than I feel inclined to summon: *"L'Encyclopédie,* dictionnaire technique," *L'Encyclopédie et le progrès des sciences et des techniques,* pp. 187-214.

12. *The Illustrations from the Works of Andreas Vesalius* (Cleveland and New York: World, 1950).

13. The best statement of the case for a social catastrophe remains the work of J. L. and B. Hammond, *The Village Labourer, The Town Labourer,* and *The Skilled Labourer* (London and New York: Longmans, 1912, 1918, and 1919). An introduction to the revisionist view will be found in T. S. Ashton, *The Industrial Revolution* (London: Home University Library, 1948), and F. A. Hayek, ed., *Capitalism and the Historians* (Chicago: University of Chicago Press, 1954).

14. The writings of John U. Nef have been fundamental in establishing the evolutionary aspect of technology. They will, moreover, be very rewarding to any reader interested in the relation of industrialization to civilisation: *Industry and Government in France and England, 1540-1640* (Philadelphia: American Philosophical Society, 1940); *War and Human Progress* (Cambridge: Harvard University Press, 1950); *La naissance de la civilisation industrielle et le monde contemporain* (Paris: Colin, 1954).

Bibliography

Scholarly literature on the *Encyclopedia,* on Diderot, and on industrialization is very voluminous, of course, and the choice of works mentioned in the notes is intended only to give entry to the main subjects. Any reader interested can proceed from any one of these writings into a vast field of study. Neither would it be appropriate to undertake anything like a full bibliography of technological history, the less so since the collaborative *A History of Technology,* ed. Charles Singer, E. J. Holmyard and A. R. Hall (Oxford: At the Clarendon Press, 1954-   ) will soon have reached the 18th century. But there are a few books on special topics which the editor has found especially helpful and well done, and which if he were a reader of this work he would be grateful to have called to his attention. (Only works in English are noted).

On agriculture

A. J. Bourde, *The Influence of England on the French Agronomes* (Cambridge: At the University Press, 1953).

Arthur Young, *Travels during the years 1787, 1788, and 1789, undertaken more particularly with a view of ascertaining the cultivation, wealth, resources, and national prosperity of the kingdom of France.* (2 V.: Dublin, 1793). A classic, of which there are many later editions.

On gardening

Nan Fairbrother, *Men and Gardens* (New York: Knopf, 1956).

On sugar

J. C. Sitterson, *Sugar Country* (Lexington: University of Kentucky Press, 1953).

On military history

> Edward Meade Earle, ed., *Makers of Modern Strategy* (Princeton: Princeton University Press, 1944), especially the chapters by Felix Gilbert, "Machiavelli: The Renaissance of the Art of War," and Henry Guerlac, "Vauban: The Impact of Science on War."

On iron and steel

> T. S. Ashton, *Iron and Steel in the Industrial Revolution* (London: 1924). Arthur Raistrick, *Dynasty of Iron Founders; the Darbys and Coalbrookdale* (London and New York: Longmans, 1953).
> *Réaumur's Memoirs on Iron and Steel,* Trans. by A. G. Sisco, Introduction and Notes by Cyril Stanley Smith (Chicago: University of Chicago Press, 1956).

On the extractive industries and metallurgy (for comparison with the 18th century):

> Agricola, *De Re Metallica,* Trans. Herbert and Lou Henry Hoover (New York: Dover, 1950).
> Vannucio Biringucci, *Pirotechnica,* Trans. and ed. Cyril Stanley Smith and Martha Teach Gnudi (New York: The American Institute of Mining and Metallurgical Engineers, 1942).

On glass

> Warren C. Scoville, *Capitalism and French Glassmaking, 1640-1789* (Berkeley: University of California Press, 1950).

On paper

> Dard Hunter, *Papermaking* (New York: Knopf, 1947).

On textiles

> Luther Hooper, *Hand Loom Weaving* (London: Pitman, 1926).

On inventions

> A. P. Usher, *A History of Mechanical Inventions* (Rev. ed., Cambridge: Harvard University Press, 1954).

On economic history

> Henri Sée, *Economic and Social Conditions in France during the Eighteenth Century* (New York: Knopf, 1927).

Finally, the editor has often turned for guidance, and seldom in vain, to that prosaic English-language successor to the *Encyclopédie*, the *Encyclopaedia Britannica*. The eleventh edition is rather more generous with information on technical history than are those which have appeared since that work came so incongruously to rest in the city of Chicago. But the essential source of information is, of course, the *Encyclopédie* itself. Where no other reference is given, quotations are drawn from the appropriate article in the original.

Plates

In this Dover edition the original Diderot plates have occasionally been separated into their parts so that they could be reproduced without reduction in size. Such separate illustrations which come from a single Diderot plate have been assigned a single Dover plate number. Their attribution in the Diderot volumes is given at the lower right hand corner.

The following abbreviations have been used to indicate location in the original set:

"Recueil de Planches sur les Sciences, les Arts Liberaux et les Arts Mechaniques avec leur Explication,"

Vol. I, from Primiere Livraison; Paris; Briasson, David, LeBreton, Durand; 1763.

Vol. II, from Seconde Livraison en Deux Parties, Premiere Partie; Paris; Briasson, David, LeBreton, Durand; 1763.

Vol. III, from Seconde Livraison en Deux Parties, Seconde Partie; Paris; Briasson, David, LeBreton, Durand; 1763.

Vol. IV, from Troisieme Livraison; Paris; Briasson, David, LeBreton; 1765.

Vol. V. from Quatrieme Livraison; Paris; Briasson, David, LeBreton; 1767.

Vol. VI, from Cinquieme Livraison ou Sixieme Volume; Paris; Briasson, David, LeBreton; 1768.

Vol. VII, from Sixieme Livraison ou Septieme Volume; Paris; Briasson; LeBreton; 1769.

Vol. VIII, from Septieme Livraison ou Huitieme Volume; Paris; Briasson; 1771.

Vol. IX, from Huitieme Livraison ou Nouvieme Volume; Paris; Briasson; 1771.

Vol. X, from Neuvieme Livraison ou Dixieme Volume; Paris; Briasson; 1772.

Vol. XI, from Dixieme et Derniere Livraison ou Onzieme et Derniere Volume; Paris; Briasson; 1772.

The plates keyed "Supplement," have been taken from the "Suite du Recueil de Planches, sur les Sciences, les Arts Liberaux, et les Arts Mechaniques, avec leur Explication" (Paris; Panckoucke, Stoupe, Brunet; 1777).

Agriculture & Rural Arts

Plate 1 Agriculture

Agriculture, wrote Diderot, is of all the arts "the first, the most useful, the most extensive, and perhaps the most essential," and another Encyclopedist describes it as "the most innocent." This profoundly French instinct still expresses itself in beautiful cultivation which gently spreads the most humane landscape in the world all across the countryside of France. The Encyclopedic spirit found it peculiarly painful, therefore, that the special virtue of agriculture should have been compromised by backward technique. Unfortunately, however, England, not France, was the home of the 18th century revolution in methods—deep plowing and even sowing, systematic crop rotation, and proper restoration of the soil—and France was slow to follow. The inattention of many landowners who preferred Paris to the provinces was compounded by the mulish conservatism of their peasants. From both ends of the social structure, feudal organization entangled cultivation in an unenlightened routine.

Closing the gulf between theory and practice was the first mission of the Encyclopedia, *and the plate illustrates farming, not as it usually was, but as it might have been on some model manor. The fortifications of the chateau fall in romantic ruins. Their menace is no more. In the village stand church and mill to render necessary services. Fields are no longer parcelled out in crabbed medieval strips, but are enclosed to permit a rational husbandry. Peasants actually use the tools constantly urged upon them by experts—implements which in real life they stubbornly refused to touch. The plow (Fig. 1) is the type invented by Jethro Tull, the English agrarian reformer and crank. Brisk horses have replaced lethargic oxen. A woman pushes a light, easily-handled seeder (Fig. 4) along the furrow the plowman has just opened. The device drops seed evenly and economically, so that the seed is immediately covered with the dirt turned out of the next furrow cut by the plow. This was a French invention; it was designed by the Abbé Soumille, one of the many 18th century priests whose services to humanity took the form of technological projects. In the background a worker sows grain, (Fig. 5), while a light spike-harrow (Fig. 6) distributes and covers the seed. A roller (Fig. 7) firms the ground. In all of these operations, it will be observed, labor is co-ordinated.*

Plate 1   Agriculture

Plate 2   Farmyard

This is a representative farmyard. It hardly differs from those which still lie behind the most prosperous facades in any of the thousands of villages in which the farmhouses of Europe cluster. The village was the farming unit in the manorial system, and this is the origin of one of the features that lends the landscape of the old world a quality different from that of America, where the practice of family farming scattered homesteads across the countryside.

The backdoor of the farmer's house (I) overlooks his outbuildings, and an archway (P) gives wagons independent access to the village street. A door (Q) leads to the root cellar, and between it and the archway are the stables and drinking troughs. Continuing around the court are a press, probably for wine (H), and across the foreground a series of emplacements for buildings which the artist leaves to the imagination in order to afford a view: a winery (G), dairy (F), gate to the fields (E), cow barn, sheepfold and duck pond, etc. (D, C, B, K). What looks like a silo is in fact a dovecote (A)—silos were not yet in use. The building at the right (M) is used for chickens, turkeys, and pigsties (L), and at the far end (O) is the carriage house.

Here is a farm, then, with a plant thoroughly equipped for all the subsidiary agricultural pursuits, some of which are illustrated in the following plates.

Plate 2   Farmyard

Vol. I, Oeconomie rustique, Basse Cour.

Plate 3   Making Hay

Vol. I, Agriculture, Façon des foins et Moisson.

Plate 3   Making Hay

Scythe men cut the crop (Fig. 1) which is raked up and stacked in small heaps by women. One of the mowers relaxes after a lunch of bread, cheese, and wine. In Figure 2, women with sickles carefully cut ripe wheat, and lay it in parallel rows. It is then tied in sheaves and carried to the barn, where it is threshed.

Plate 4 Threshing

Plate 4 Threshing

Sheaves are piled just inside the barn door (Fig. 1), to be opened and spread about the floor for threshing (Fig. 3). Here the 18th century revolution in agricultural methods stopped. The flails wielded by the threshers do not differ much from those in use 500 years earlier. After the grain has been beaten from the heads, it is shovelled into a pile, and scooped up in a winnowing basket which separates the chaff. The artist does not show the dust which fills the air and chokes the threshers. Threshing was notoriously the hardest and most disagreeable work on the farm.

Plate 5   Gardening I

In the classical French view gardening imposes geometric decency on a nature not to be left to its own slovenly instincts and unkempt effects. This is a feeling which has permeated all levels of society. Rows of severely cropped trees stalk squarely down the poorest village street, and on a larger scale they sweep grandly up the terrace at Saint-Germain-en-Laye and along the prospects of Versailles. There is no more fitting symbol of French gardening than the pruning hook.

Plate 5 Gardening I

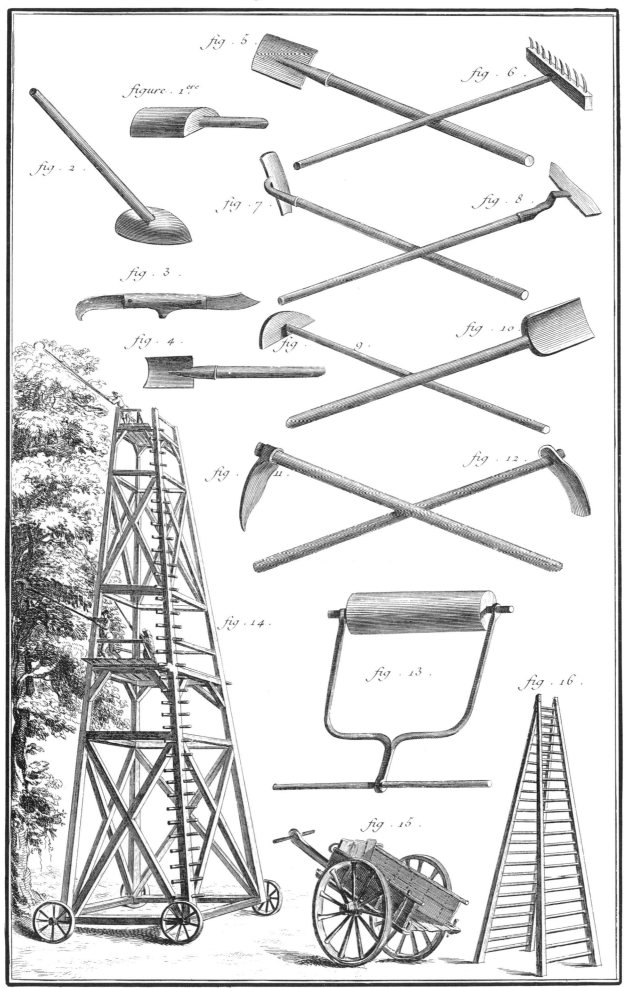

fig . 5 .

fig . 6 .

figure . 1ᵉʳᵉ

fig . 2 .

fig . 7 .

fig . 8 .

fig . 3 .

fig . 4 .

fig . 9 .

fig . 10 .

fig . 11 .

fig . 12 .

fig . 14 .

fig . 13 .

fig . 16 .

fig . 15 .

Plates 6, 7, 8, 9   Gardening II, III, IV, V

In Men and Gardens, *a recent and very charming historical essay, Nan Fairbrother entitles her chapter on the French grand manner "Les jardins de l'intelligence—The Gardens of the Mind." In the work of LeNôtre, the chief gardener of the Sun King Louis XIV, the spirit of French classicism moved out of doors to express its Cartesian vision of order in compositions of boxwood, yew, and poplar, of lawn, pool, and fountain.*

In his own specialty, LeNôtre was the peer of Corneille, Molière and Racine, the great dramatists who made of the French language an instrument of incomparable clarity and precision, or of the Marshals of France, Turenne and Vauban, who regularized the arts of warfare. In LeNôtre's mind gardening was essentially an architectural art, which required that arrangements pleasing to the intelligence must be imposed on otherwise meaningless space. According to his principles, designs should not be predictable, for then they would be banal, but once understood, they must be seen as necessary. He speaks, not in pretty flowers to the romantic fancy, but rather to the rational mind and in the language of geometry. In this conception of gardening, blossoms have little place. They will be employed only as masses of color to be manipulated, but very sparingly, since they are all too likely to strike a jarring note. On the whole, LeNôtre preferred to work with lawns and pools and gravel paths.

His work has enjoyed a permanence granted to few gardeners. It may still be appreciated at Versailles and in the Tuileries, the original plan for which is given in Plate 8, Fig. 1. Any tourist who gazes from the Louvre across the Tuileries and up the incomparable vista of the Champs Elysées is admiring a prospect opened to him by LeNôtre. Some of LeNôtre's designs for smaller gardens will be found in the plates that follow. Gardeners who wish to adopt them will find a scale at the bottom (the toise *is a little over six feet). But they must be warned that LeNôtre was very severe about sizes and areas. A really elegant garden could not be cramped into less than 30 or 40 acres. But in as little as 5 or 6 acres it might be just possible to create a not uninteresting small garden.*

Plate 6   Gardening II

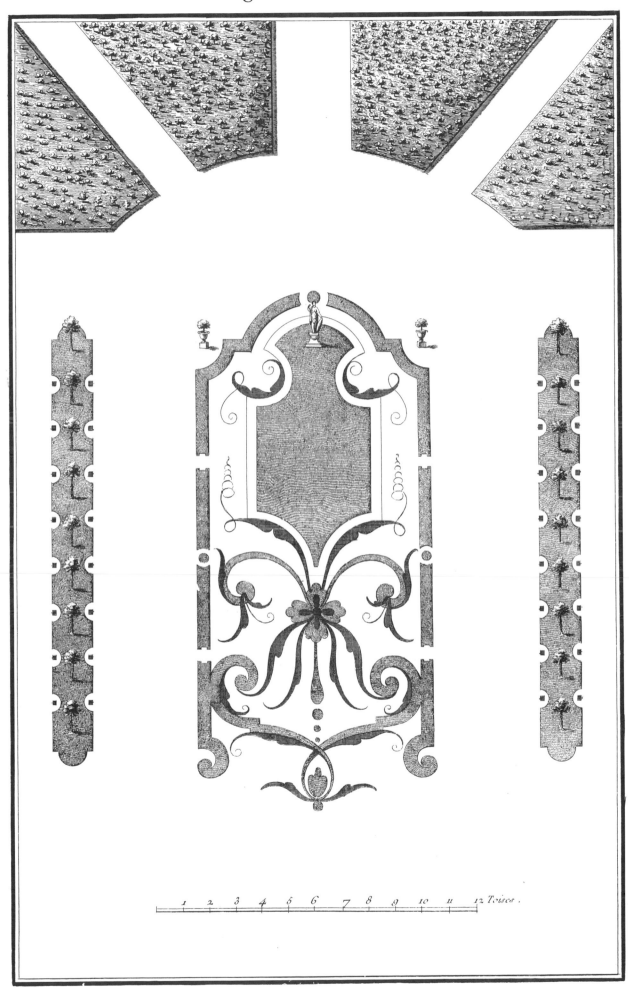

1 2 3 4 5 6 7 8 9 10 11 12 *Toises*.

Plate 7   Gardening III

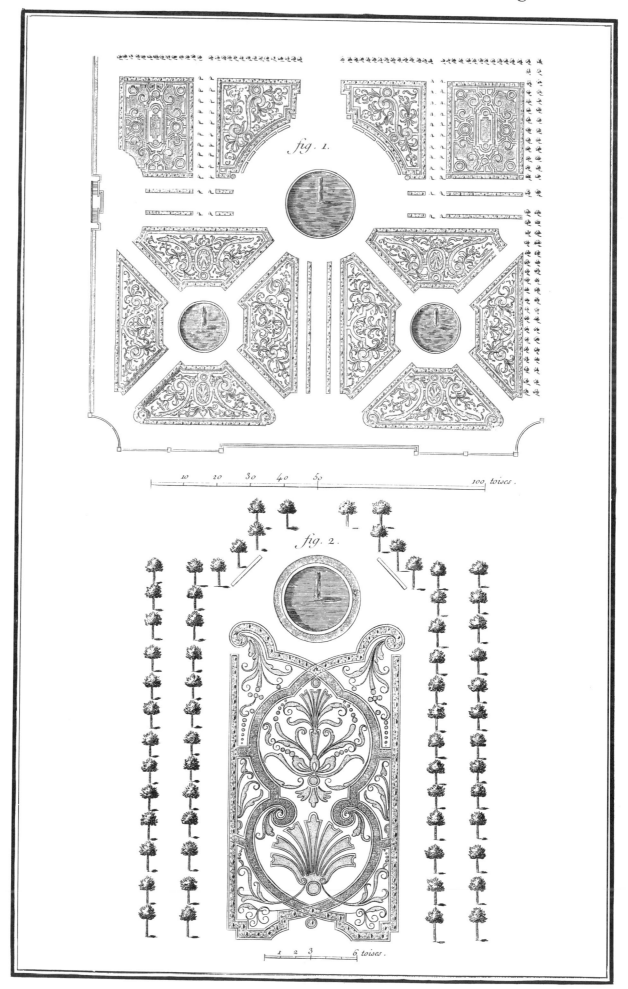

fig. 1.

10   20   30   40   50   100 toises.

fig. 2.

1 2 3   6 toises.

Plate 8 Gardening IV

1 2 3 4 5   10   15   20 Toises.

Plate 9   Gardening V

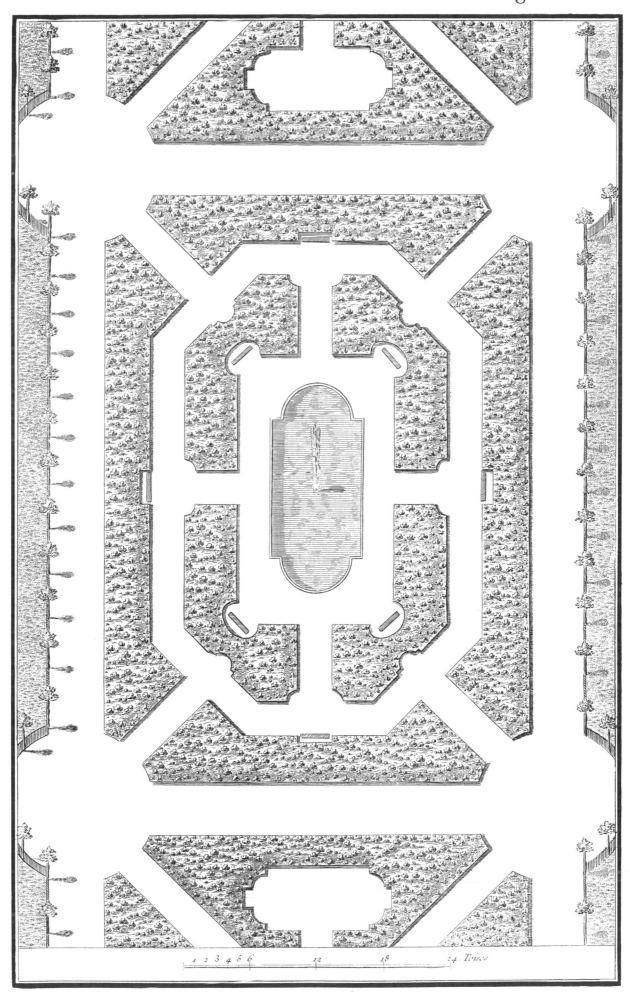

1 2 3 4 5 6   12   18   24 Toises

Plate 10   Gardening VI

Vol. I, Agriculture, Jardin potager.

Plate 10   Gardening VI

This plate represents a specialty garden. Emphasis is upon perfection of fruit and vegetable, rather than upon quantity of produce. Espaliered trees line the walls. A well provides water for the small beds. D^1 is a bed which has been deep-trenched, and D^2 one which is planted with vegetables under cloches. The last two rows are slanted to catch maximum sunlight. Small plants like strawberries would be grown under such shelter. This form of gardening is not generally practiced in the United States; cloches are still extensively used in Great Britain and France.

Plate 11   Butter

Vol. I. Oeconomie rustique, Laiterie.

Plate 11   Butter

A milkmaid works her churn. This, says the Encyclopedist, is a dairy on one of the royal estates: "Hence it is somewhat more elegant than the average." The machine at the right (Fig. 3) is a Flemish butter churn, which the authors consider superior to the conventional type (Fig. 1).

Plate 12   Cheese I

From the myriad variety of French cheeses, the Encyclopedia *chooses two to illustrate, both from mountain regions: cantal from Auvergne, and gruyère, manufactured in the Vosges and the Jura, as well as Alpine Switzerland. The vignette shows the interior of one of the thatched stone cottages which dotted the mountain meadows sloping up to the forbidding puys or peaks of the Massif Central, where cattle are pastured from mid-May to mid-October. This was the cheese-making season, and cheese-making was the chief occupation of the dark, heavy-countenanced Auvergnat herdsman.*

Plate 12 Cheese I

It would be tedious to go into detail. Suffice it to say that what gives distinctive character to cantal *is that it is a pressed cheese (Fig. 10 shows the whey being extruded), and that its rennet—or starting culture—is made of milk from the stomach of a sucking calf (fermented in the stomach itself) together with dried pieces of stomach lining from steers, goats, and rams.*

Desmarets, who investigated the dairy industry, was aware that there is no correlation between the cleanliness of kitchens and aprons and the quality of cheeses. But this is a demoralizing fact. Like most Encyclopedists he had a tidy mind and was shocked by the filthy state in which Auvergnat herders kept their beasts and their dairies, "where there reigns an uncleanliness which it would be impossible to make too great an effort to overcome."

Plate 13   Cheese II

As "Swiss cheese" gruyère is more familiar to Americans than cantal, and was, indeed, more highly esteemed in France. Apparently Swiss cleanliness was already legendary in the 18th century, for Desmarets could not praise it too highly—though perhaps one should say Alsatian cleanliness, for the cottage here illustrated was in the Vosges. But if he was pleased with the conditions of manufacture, he was offended by the contrast these presented with the herdsmen themselves, mostly Germanic Anabaptists, who went unshaven for religious reasons and presented a most shaggy and "disgusting" appearance. Out of deference, perhaps, to the sensibilities of French readers, the artist has reduced the beard of the dairyman at the right to civilized proportions.

What distinguished gruyère was not such exotic ingredients as those of cantal, but long and careful cooking. And indeed Swiss cheese slides cleanly over the palate, tasting of some bright and antiseptic purification rather than of flavorful corruption and dark fermentings better savored than analyzed.

Plate 13   Cheese II

Plate 14   Beekeeping

Beekeeping was widespread in rural Europe. The best honey in France came from the southern provinces of Narbonne and Roussillon, where there were countless peasant establishments like the one shown here.

Different sorts of hives are shown under the shelter at the left: wood (1 and 4), osiers (2), plaited straw (3), glass (5). Number 6 is the most primitive form: a hollow tree trunk with a straw bonnet. The expert who wrote for the Encyclopedia *seems to have favored the one in the foreground (7), which has a clay base and a straw bonnet.*

Various activities are shown. The peasant in 8 jolts a swarm of bees from the upper hive into an empty one turned upside down to receive them. The two men at 10 are trying to capture an escaped swarm. One shakes the branch, while the other holds up a box into which the bees may light. The two men at 9 are beating cymbals or pans to frighten off a raiding swarm.

"There is still much to learn about bees," wrote the Encyclopedist. "Many traits of their industry and particularly of their sentiments *still elude the patience and sagacity of observers."*

Plate 13   Cheese II

Fig. 1.

Fig. 2.

Fig. 3.

Fig. 13 C

Fig. 15

Fig. 6 b

Fig. 4   Fig. 5

Fig. 5.

Plate 14   Beekeeping

Beekeeping was widespread in rural Europe. The best honey in France came from the southern provinces of Narbonne and Roussillon, where there were countless peasant establishments like the one shown here.

Different sorts of hives are shown under the shelter at the left: wood (1 and 4), osiers (2), plaited straw (3), glass (5). Number 6 is the most primitive form: a hollow tree trunk with a straw bonnet. The expert who wrote for the Encyclopedia *seems to have favored the one in the foreground (7), which has a clay base and a straw bonnet.*

Various activities are shown. The peasant in 8 jolts a swarm of bees from the upper hive into an empty one turned upside down to receive them. The two men at 10 are trying to capture an escaped swarm. One shakes the branch, while the other holds up a box into which the bees may light. The two men at 9 are beating cymbals or pans to frighten off a raiding swarm.

"There is still much to learn about bees," wrote the Encyclopedist. "Many traits of their industry and particularly of their sentiments *still elude the patience and sagacity of observers."*

Plate 14   Beekeeping

Plate 15

Plate 15   Silkworms

The Encyclopedia's *term for the cultivation of silkworms was "Education." This industry, too, belongs to the south. The newly-hatched worms are fed on mulberry leaves (Fig. 5)—they refuse to touch any but the freshest—and are brought up in trays (Figs. 1, 3, 4) in which they wriggle about until they are ready to spin their silken cocoons on twigs prepared for them (a, b, c, d). They are delicate but voracious creatures requiring a great deal of attention. The attendant at the left is separating sick and languishing worms from the robust ones "which seem disposed to make silk." They express this disposition by raising their foreparts and waving them about. Feeding lasts 42 days, during which the worm falls asleep four times to awaken in a new and larger skin on each occasion.*

The stages in a silkworm's life are shown facing. Fig. 6 is a hatching tray, in which some of the eggs have hatched. The feeding and growth of the worm are illustrated in Figs. 7, 8, and 9. In Fig. 10 it spins its cocoon. For convenience in handling, it is sometimes furnished with a curl of paper (Fig. 11) instead of a twig. Fig. 12 is a completed cocoon, from which the fluff is removed (Fig. 13). The chrysalis inside must now be killed without damaging the silk. This is accomplished by suffocation. But a few chrysalises are set aside to hatch out into moths which will lay eggs for the next crop. In Fig. 14, the moth (17 and 18) breaks through the cocoon. Fig. 15 is the chrysalis, and 16 is a cross-section of the cocoon showing the chrysalis within. For the manufacture of silk, see Plates 315-321.

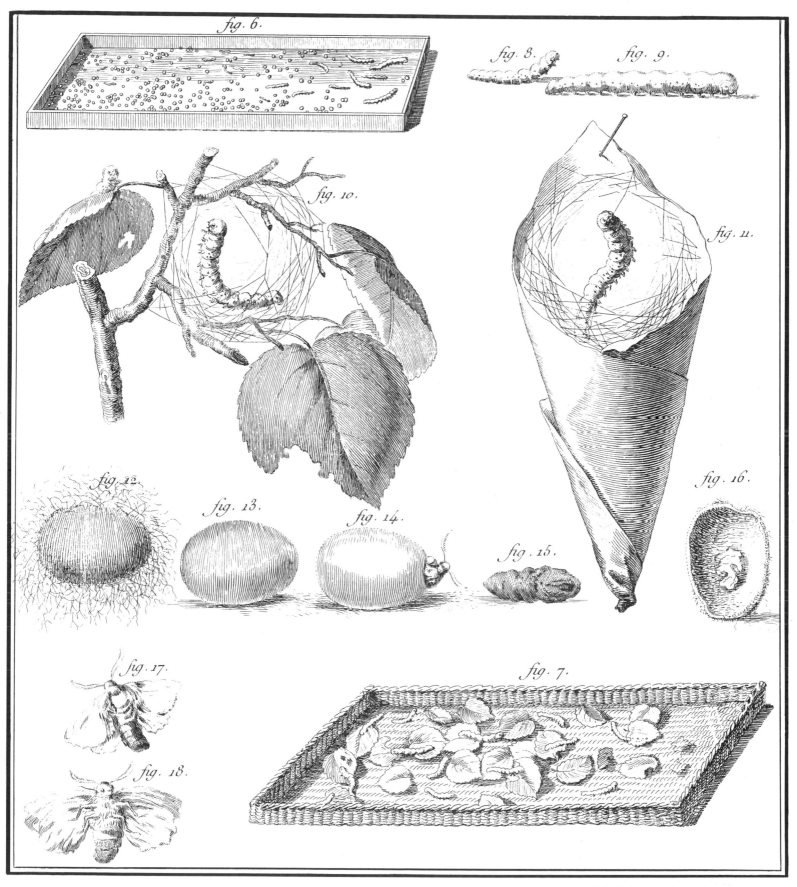

fig. 6.

fig. 8. *fig. 9.*

fig. 10.

fig. 11.

fig. 12.

fig. 13.

fig. 14.

fig. 15.

fig. 16.

fig. 17.

fig. 18.

fig. 7.

Plate 16   Windmills I

Before the Industrial Revolution, power was a rural product, drawn directly from nature. Until the development of James Watt's steam engine at the end of the century, man was still dependent upon draught animals, manpower and the forces of wind and water.

This is a wind mill, a post mill such as might have been found in rural regions throughout western Europe. The two upper arms are left unsailed to show their construction. This sort of mill differs from the more familiar Dutch wind mill in that the whole mill revolves on a pivot, whereas in the Dutch mill the building remains stationary, while a cap revolves.

Plate 16   Windmills I

Plate 17   Windmills II

Drawing advantage from wind and waterpower required mechanical ingenuity of a very high order, as will be apparent from this plate illustrating the internal construction of a windmill. To point out only the main features, the whole mill turns on the pivot (B) to face the wind. The speed can be regulated by a brake (65) on the great brake-wheel. Grain is fed through a hopper (72) into the stone millwheel (66), which is served from a platform called the "donkey's back" at the heart of the mill.

Plate 17   Windmills II

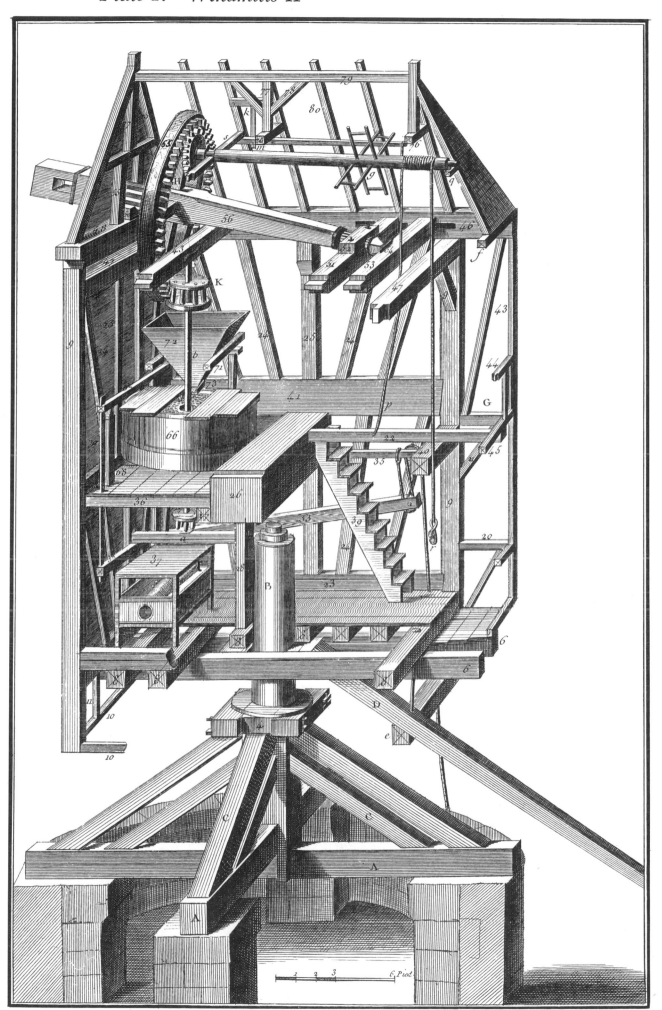

Plate 18   Milling I

Watermills were more reliable than windmills, and could deliver more power. This one is arranged to drive a millwheel (K), again for grinding flour. But watermills could also be used to run machinery, as they were in the earliest textile factories which, for that reason, were built in country regions next to falls of water. The very phrases "cotton mill" or "steel mill" are relics of the days when industry was powered by waterwheels.

Plate 18   Milling I

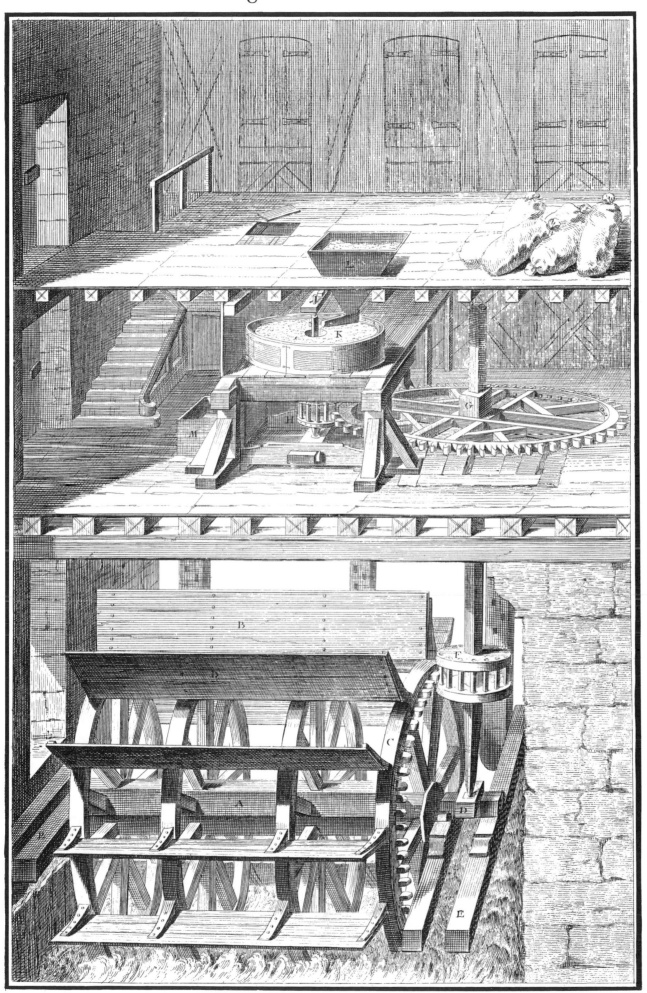

Plate 19   Milling II

Rivière des Nonnettes

B L Prevost del

Touses

Supplement, Meunier, Pl. I.

Plate 19   Milling II

A peasant brings his sacks of grain to the miller to be ground. The mill is equipped with a pulley and hoist which will pick the sack right off the horse or mule and lift it straight to the top of the mill.

Plates 20, 21   Milling III & IV

Two cross-sections through the mill, one from the side and one from the front, illustrate the milling of flour, from the hopper (Pl. 20, Fig. 12) through the bolter or rotary sifter to the grindstone (Pl. 20, V), the outlet of which appears more clearly in the frontal section of Pl. 21, i.

Plate 20 Milling III

Coupe sur la largeur

Echelle de 4 Toises.

Toises

1   2   3   4

Plate 21 Milling IV

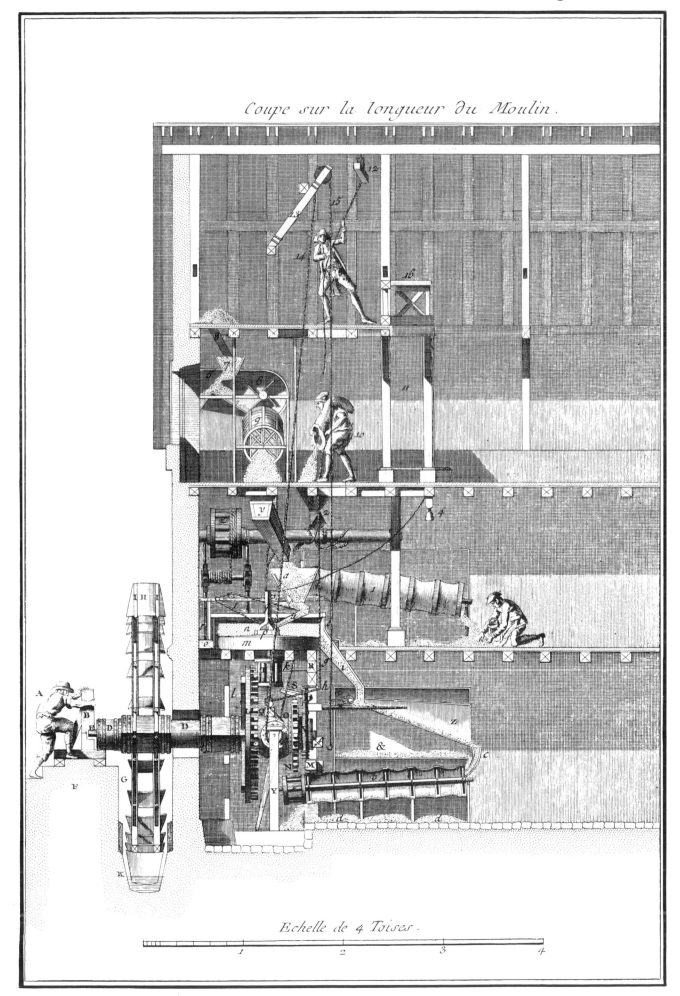

Coupe sur la longueur du Moulin.

Echelle de 4 Toises.

Plate 22   Wine

The limited availability of power put a premium upon straightforward mechanical ingenuity. Here is a double-action wine press, invented by the Abbé Legros, curate of Marsaux, "a man born for the mathematical arts." It can be worked by one man, and it saves time: while the grapes in the right-hand chamber are being crushed by approaching walls, the already pressed pulp in the left-hand chamber is draining into a tub. A siphon transfers the must to a larger tub, from which it is forced, by a primitive pump, into the row of barrels. Successive siphonings enable the lees to settle.

This picture, like that of the model farm, represents a highly idealized operation. One of its purposes is the exclusion of air. "Everybody knows that air and lees are the enemies of wine," wrote another technically minded priest, the Abbé Pluche.

Plate 22

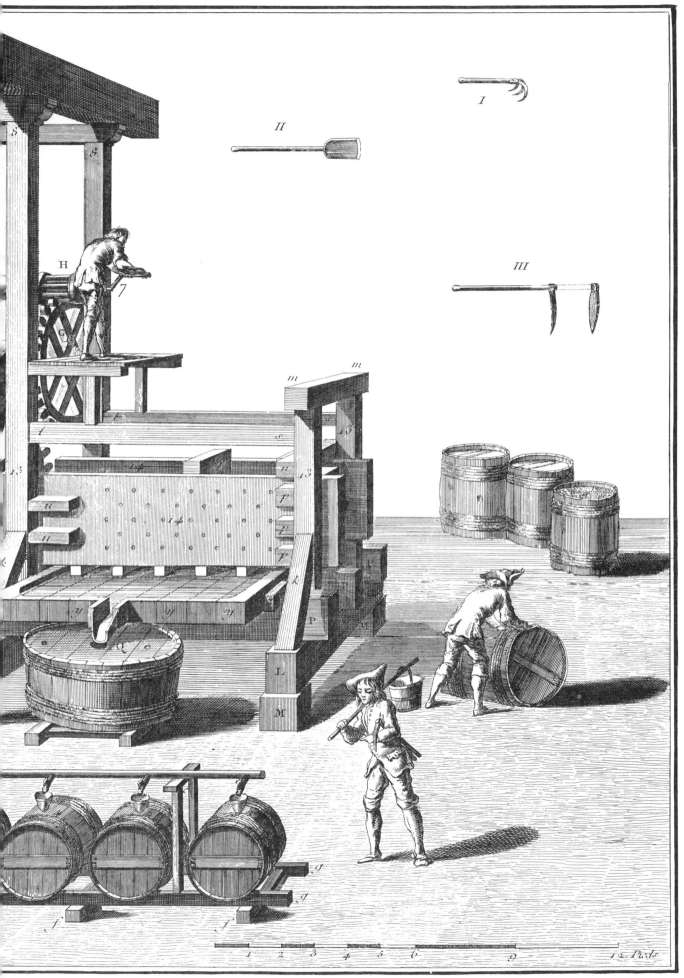

Plate 23

fig. 1

Plate 23 Cider

Unlike the Abbé Legros' wine press, which in all probability never got off the drawing board, this cider press has an air of the farm. Normandy was orchard country. There apples took the place of grapes, cider of wine, and applejack (calvados) of wine-brandies like cognac and armagnac. The apples are tossed into compartments according to grade (T, L, etc.), and crushed in the circular trough (R, P) surrounding the bins. A horse is hitched to the stone wheel. Then, cider is simply squeezed out of the mass of crushed apples (F) in the great press at the left.

Cider

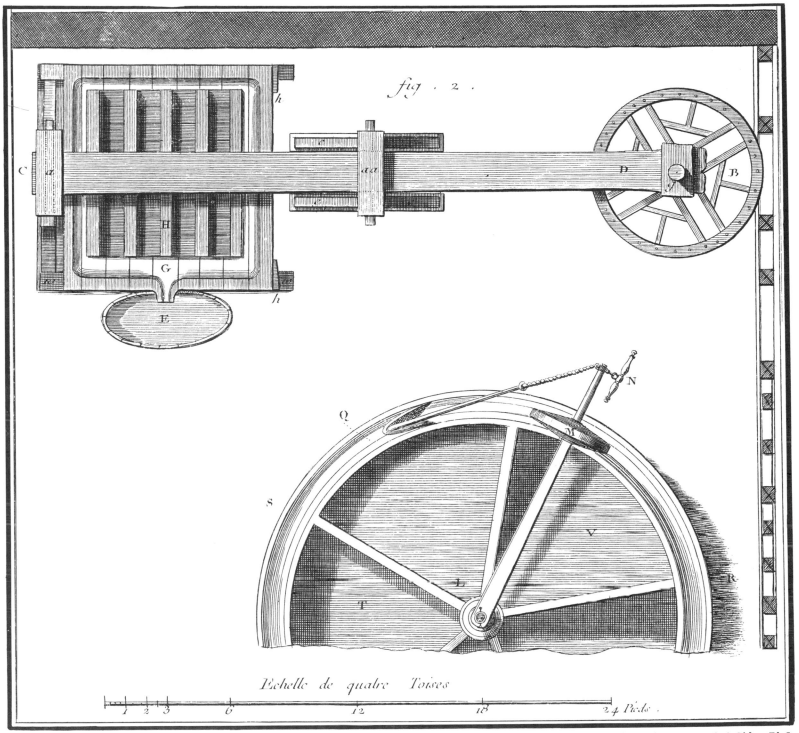

Vol. I, Oeconomie rustique, Pressoir à Cidre, Pl. I.

Plate 24   Charcoal I

Next to mining, charcoal-burning was probably the most storied industry of rural Europe. Its practitioners lived their own sooty lives in the woods, apart from settled communities.

This plate illustrates slightly different ways of building a charcoal furnace. In one case the wood is piled around a central pole; in the other a passage is left through the center of the heap. These differences, however, are trivial. In all cases the furnace is covered with a clay made of earth and powdered charcoal dust, a process known as "putting on the shirt" (Fig. 3). This layer of earth holds combustion down to the minimum level so that the product will be charcoal instead of ashes.

Plate 24   Charcoal I

Plate 25   Charcoal II

Plate 25   Charcoal II

A workman (Fig. 4) lights a furnace through the top, and the combustion gets under way (Fig. 5). As it proceeds more air is needed, and vents are opened in the "chemise" (Fig. 6). The fire must be tended constantly to regulate the rate (7, 8), until at the end the great cone of wood is a small heap of charcoal (11).

Plate 26 Making Wooden Shoes

Vol. I, Oeconomie rustique, Manière de faire les Sabots.

Plate 26   Making Wooden Shoes

Sabots *remain the footgear of many working men in rural France, and may occasion-*
ally be heard clattering through the streets of Paris. In the 18th century, they were
worn by nearly all the peasants, and were as much the mark of a Frenchman as of a
Dutchman. In England, they were unknown. The Encyclopedia *tells of an English*
physician who ordered wooden shoes for a boy who was suffering from rickets; the
prescription could be filled only by sending to France.

Manufacture of wooden shoes was simple and quick. A block of wood—usually aspen
—is trimmed roughly to shape by the adze (Fig. 1); hollowed out by various-sized
spoon augers (Figs. 2 and 3); and given a final trim with a bench-cutter (Fig. 4). The
stakes at the right (Fig. 5) are a by-product to be used for supporting grapevines.

Plate 27   Preparing Hemp I

Plate 27   Preparing Hemp I

Preparing hemp was an occupation suited to family groups, and therefore to peasant industry. First, the plant (a cool weather annual) was harvested. The male plants were pulled out, roots and all, allowed to dry, and beaten against a wall or tree to remove the dried leaves and blossoms. This process was so simple that it is not shown.

The female plant had to be treated differently. It was allowed to remain in the ground two weeks longer than the male, so that the seeds might ripen. The plants, after drying, were pulled through combs (r) to remove the leaves and seed. The finest seed was kept for the next crop, and the rest pressed for oil or sold as poultry feed. The riper female plant produced a coarser fiber than the male.

From this point on, both sexes of hemp were processed alike. Reduced to bare skeletons, the stalks were steeped or "retted" in a ditch full of stagnant water. There they were weighted down under the scummy surface by planks loaded with rocks (q). Retting softened the bark, from which fiber was spun, and allowed it

Plate 28   Preparing Hemp II

to be stripped from the stem (Fig. 4) after drying. This was an easy task, fit for a child, but some families preferred to dry the plants in an oven (Fig. 3) fueled by hemp stalks and tended by an old but attentive woman ("I say an attentive woman," writes the Encyclopedist severely, for the hemp must not be scorched), and then to crush the bark off the stems in a special machine (U). Finally (Fig. 6-8), the crude hemp was twisted into queues, beaten with paddles against a board to knock out all remnants of stem, and gathered into bundles (P) to send to the factory.

Plate 28   Preparing Hemp II

At the factory, hemp fiber is stripped through five successive carding combs (Figures 1, 2, 3), pulled around a bar (A), and rubbed over a rough surface (R). These operations remove unwanted plant tissue, and prepare it to be spun into rope or thread. The Encyclopedists, however, do not follow it out of the domain of agriculture and "rural economy" into manufacturing.

Plate 29 Tobacco I

In the Encyclopedists' classification of human activity, not only farming but the production of commodities drawn from plants belongs to "rustic economy". Tobacco is the first of the agricultural industries to be illustrated. Under the paternalistic economic policy of the old regime, the cultivation of the tobacco plant was forbidden in France itself. Partly, this was to protect the planters of the French West Indies. But more important, the manufacture and distribution of tobacco was farmed out to a privileged monopoly from which the state received in return a very considerable revenue. Governments have always found tobacco highly taxable.

Tobacco leaves were imported in barrels, from which they have been unpacked in the upper vignette, where one workman (Fig. 1) is removing the leaves damaged in transit. His fellow (Fig. 2) is pulling from a mass of tobacco (E) the bundles or swatches in which it was tied to be dried after picking. He grades and sorts them into bins according to quality.

In the lower vignette, a workman picks out the best and biggest leaves to be used as wrappers, after which he will carefully moisten them in the tub at the rear. The bulk of the tobacco is piled at the right, where it too is humidified (Fig. 2) so that it can be worked up into ropes.

Plate 29 Tobacco I

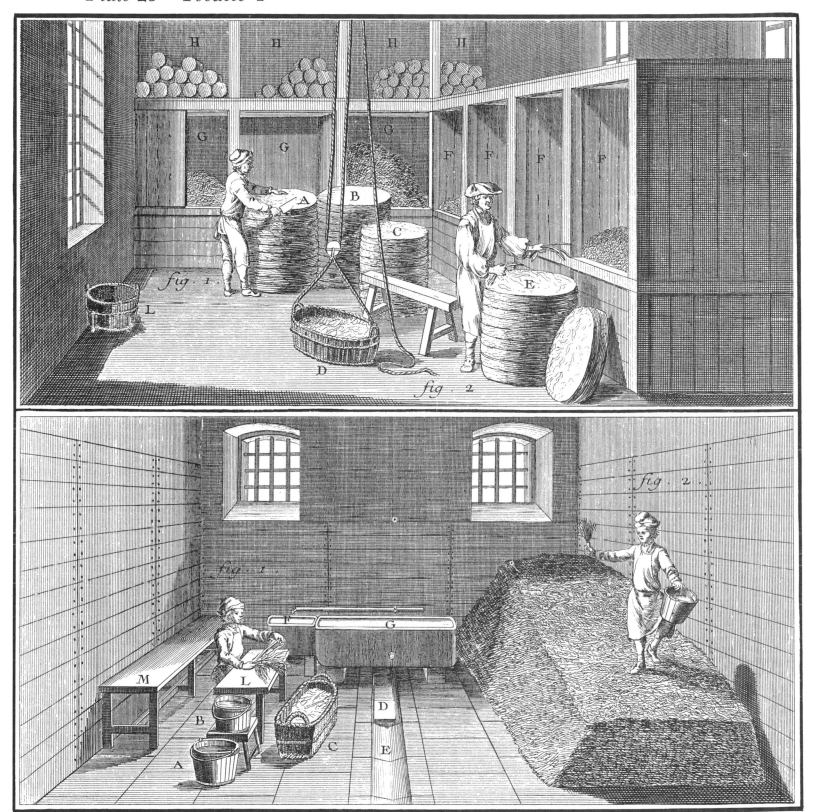

Plate 30 Tobacco II

A pulley hoists the humidified tobacco into the stripping room (upper vignette), where boys remove the fibrous ribs from the leaves.

The lower vignette shows two different methods for twisting the leaves. In the French method, at the right, the work is done entirely by hand. Each "spinner" is served by two children sitting below his bench (only one of each pair is shown). One hands him a bunch of leaves to which the apprentice has given a preliminary twist. The spinner fashions it into a rope and is then supplied with a wrapped leaf prepared by his other apprentice. In the "Dutch method" at the left a reel works the leaves into a rope.

Plate 30 Tobacco II

Plate 31   Tobacco III

Tobacco is drawn off a reel (Fig. 1) and wound around a bobbin (Fig. 2) into a tight roll several layers thick. The rolls are then compressed (Fig. 4), and—though this is not shown — closely wrapped and aged in a warehouse until fit to be sold as pipe tobacco.

Snuff, the other main product of the industry, required further pressing. First, the roll is cut into lengths (below, Fig. 1), which are sorted according to size (Fig. 2).

Plate 31   Tobacco III

Plate 32   Tobacco IV

Snuff was far more closely compressed than pipe tobacco. Six or eight of the lengths cut in the previous plates are crushed into a single bar of snuff — a "carrot" — in molds (D) made to fit into the presses here shown. There are twelve or fourteen molds in each of the five racks of the press (C). French snuff, "black rappee," dominated the trade in the eighteenth century, and its success is still commemorated in the symbolic red "carrot" of snuff which advertises that a café is also a tabac — i.e., licensed to sell tobacco. For despite the Encyclopedists, despite even the French Revolution, the sale of tobacco remains a government monopoly in the Fourth Republic, as no doubt it will continue to be in the Fifth.

Plate 32   Tobacco IV

Plate 33   Tobacco V

Snuff is packaged in a string jacket (above). The ends are neatly pared to present a smooth appearance (below). The snuff-taker, then, will have a grater with his snuff box — these were often very ornate accessories — to scrape off finely powdered snuff which he inhales by sniffing.

According to the Encyclopedia, *one of the advantages of the tobacco industry was that it did not require either complicated machinery or unusually intelligent workmen. Believers in a free economy, the Encyclopedists were indignant that so promising an industry should be confided to the stifling management of a monopoly.*

Plate 33   Tobacco V

Plate 34   Cotton I

The vignette at the top might almost be a set for Rameau's Les Indes Galantes—The Romantic Indies, *as staged at the Paris opera in 1954 and 1955. It represents a cotton plantation on which the happy slaves are aborigines depicted rather in the manner of Bernardin de St.-Pierre than of Harriet Beecher Stowe. Cotton grows as in the tropics on treelike shrubs, and not as the annual plant of the American South. A lightly-clad slave (2) plucks the bolls. His fellow (3) picks over the cotton and passes it to a woman who mills out the seed (4). Next, the lint is trampled into a bale (5) and moistened (6) to make it pack well. Finally, the bales are numbered in sequence (7) and shipped off (8) to Martinique or Guadaloupe where they will be transshipped in sea-going vessels to supply the cotton industry of Rouen and Troyes.*

In including the curious picture below, the Encyclopedists must have been yielding to the temptation of the picturesque. For it depicts the dressing of cotton by a bow, a method used in the Far East — hence the Chinese with his pipe. A bowstring is vibrated through a pile of raw cotton. One end is fastened to a laminated spring (BC), the other to the bow itself (E), and the cord is passed over a spool in the operator's right hand, with which he adjusts the tension as he works the bow. The process is laborious and slow, but the advantage is that the fibres are simply shaken apart and are not torn or bruised as in carding. In Europe bowing, though not used for cotton, was employed for dressing certain types of wool.

Plate 34   Cotton I

Plate 35 Cotton II

Carding was the technique used in Europe for separating cotton fibers. The plate is a close-up of combing cotton by hand (Figs. 1-4). After combing, something of the lustre (which is never lost in the bowing process) may be restored by twisting the fibres tightly together (Figs. 5-6) so that they polish each other.

Plate 35 Cotton II

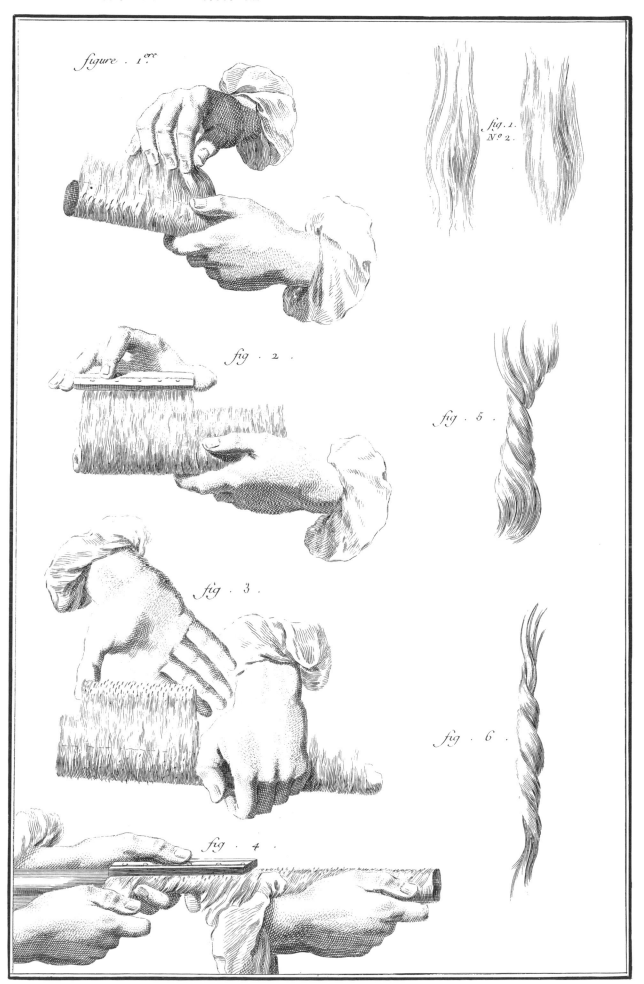

Plate 36

Plate 36   Indigo and Manioc

An indigo plantation in the West Indies is illustrated in the vignette above.

The dye is extracted from the plant by fresh water in a series of leaching basins. At the top is a reservoir (A). Plants are harvested in the fields (N) early in the morning, and a slave (I) brings them in and dumps them in the upper basin (B), where they are steeped in the heat of the sun for anywhere up to 24 hours, depending on the temperature of the day. This extracts the "indicam" or starch from the leaves. The liquor, still colorless, is then run into the lower basin (C), where it is vigorously agitated until deep blue crystals of indigo appear. The foreman verifies the progress of this oxidation by taking samples in a silver ewer (Fig. 3, facing). When crystallization is complete, the indigo is separated out and drained by means of filtering cones "like a monk's cowl" (G), and the crystalline indigo thoroughly dried in trays (Fig. 1 below) in an open shed (H). Vile-smelling fumes arise from the process, particularly from the discard basin (D) — Le diablotin — devil's tank. Long exposure

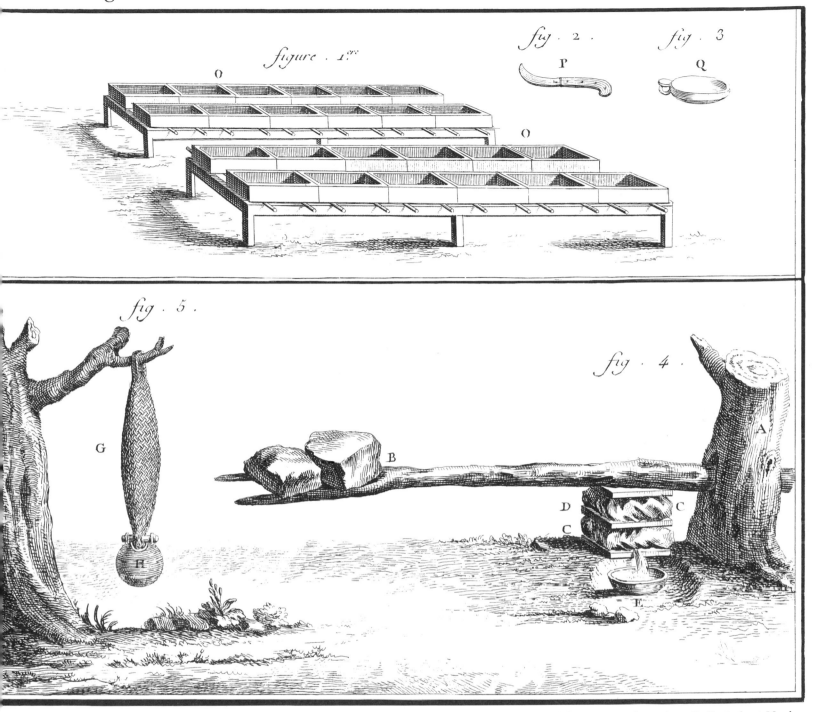

to these vapors was usually fatal to the workers. This tended to discourage expansion of the industry.

An altogether more innocuous product and primitive process is illustrated above in the pressing of manioc or cassava. The method is that of the Carib Indians, as are — evidently — the tools. In Fig. 4 cassava roots are pressed for juice, and in Fig. 5, the pulp is squeezed dry in a sort of sock weighted down by a stone. The dried residue, often called arrowroot, is the starch of which tapioca is made.

Plate 37   Sugar I

Plate 37   Sugar I

A sugar plantation, probably in Haiti, disposed according to the dictates of reason and a slave economy. The main house (1) is at a suitable distance from the slave quarters (2). Farm animals are pastured in the field in the foreground (3), and savannahs planted to sugar cane stretch up the valley and along the slope (5). The plantation does its own refining in the buildings at the left (7, 11). Power is supplied by a water wheel in the mill house (6). The millrace (9) empties into a fish pond. Areas which are too rough for sugar cane (13) may be used for cassava, bananas, coconuts, or garden vegetables. All in all, the arrangements might be those of a well-ordered plantation of the Old South.

Plate 38   Sugar II

The juice is pressed out of sugar cane by feeding it between vertical rollers. The mill in Fig. 1 is run by horsepower and in Fig. 2 by water power.

Plate 38   Sugar II

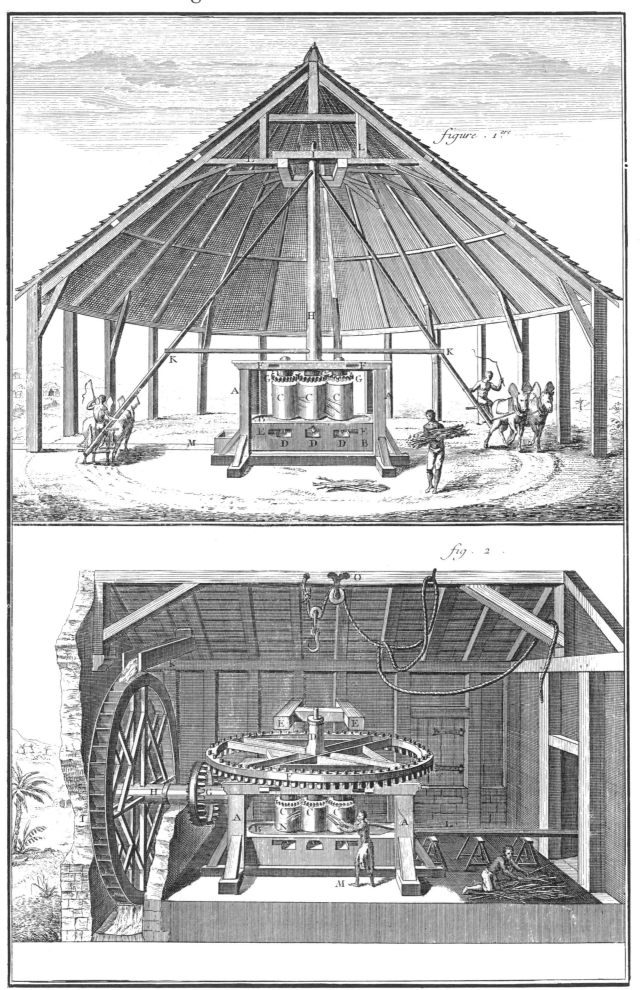

figure . 1.ere

fig . 2 .

Plate 39   Sugar III

Plate 39   Sugar III

Juice is boiled down into sugar in five open kettles, each smaller than the last, named (from left to right) the grande, *the* propre, *the* flambeau, *the* syrop, *and the* batterie. *"However we may smile at these names," writes an English commentator on the process, "so characteristic of the fanciful distinctions of the French, we think we must avail ourselves of them in rapidly pursuing the divers in and outpourings, as we should otherwise be obliged to take a very round-about way in making ourselves intelligible."[1] Juice from the receiver B is run into the* grande *where slaked lime is added. As the temperature rises, a greenish scum forms and is skimmed off by the attendant (E). Clarification progresses from one vessel to the next. The rate of boiling is higher in the* propre *than in the* grande, *in the* flambeau *than in the* propre, *and so on. Each is filled, when about half empty, from its predecessor. Finally the syrup in the* batterie *is thick enough to sugar when tested in droplets, and it is then ready for "striking" or granulation. It is ladled into coolers, shallow wooden tanks (not shown), where it crystallizes.*

[1] On the Nature and Properties of Sugar Cane *by* G. R. Porter *(Philadelphia, 1831).*

Plate 40   Sugar IV

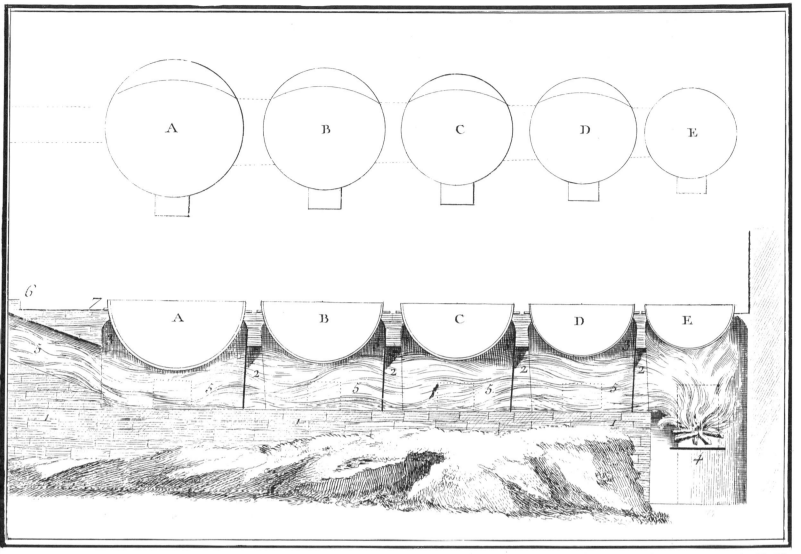

Plate 40   Sugar IV

The arrangement of the boilers.

The grande (A) is at the left and the batterie (E) directly over the fire. The intensity of the heat, therefore, increases as the evaporation continues and the size of the vessels is proportionately diminished.

Entering the United States by way of the French colony of Louisiana, the methods and even the terminology of the planters of the French West Indies exerted a profound influence over the sugar technology of the Old South. But in Haiti itself, the French sugar plantations were destroyed and the planters massacred in the slave rising of 1804, led by Jean Jacques Dessalines, the successor to the famous Haitian patriot Toussaint L'Ouverture. With Haitian independence, the old Creole culture disappeared.

Plate 41

Plate 41   Sugar V

In the British colonies, granulated sugar was "potted" and cured or drained of its molasses in hogsheads. The French preferred the process illustrated here, known as "claying." After crystallizing, the sugar is packed in conical forms of earthenware, each made with a bunghole at the apex. The forms fit point-down in pots into which the molasses is allowed to drain. This takes about 24 hours, after which the sugar is loosened somewhat (Fig. 1) and a suspension of fuller's earth, or other clay, is poured through it. This acts as a liquid sponge which takes up the remaining syrup adhering to the sugar crystals. Claying may be repeated three or four times, after which the sugar loaves are removed from the forms and dried in the sun.

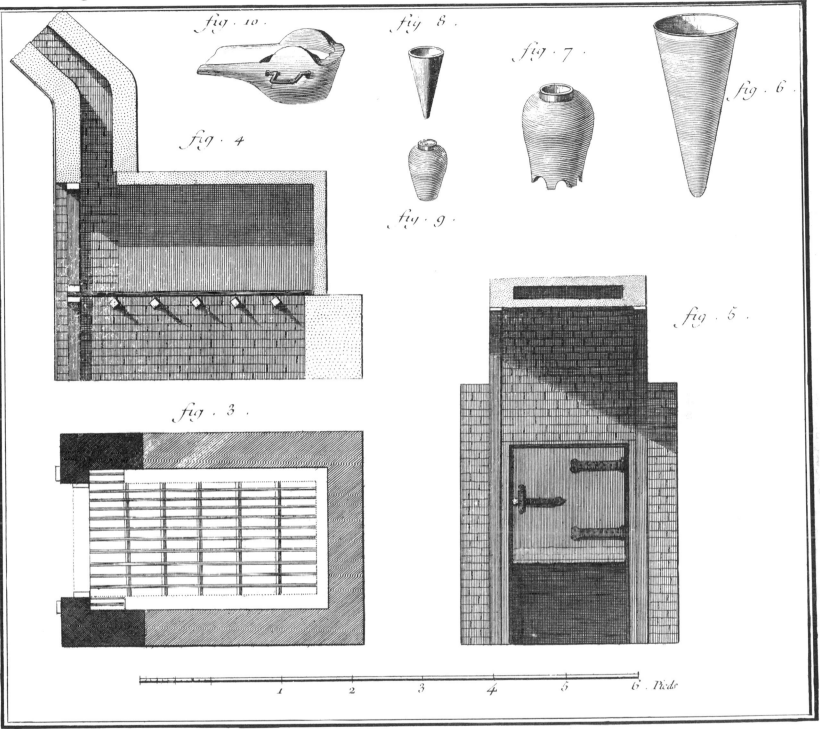

fig. 10.

fig. 8.

fig. 7.

fig. 6.

fig. 4.

fig. 9.

fig. 5.

fig. 3.

1   2   3   4   5   6 . Pieds

Plate 42

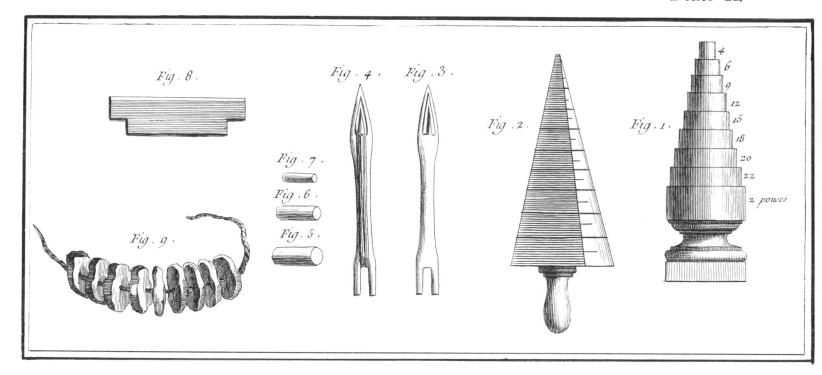

Plate 42 Fishing I

Although Diderot did not put fishing under Oeconomie rustique, *the French language gives a certain authority to do so, for its term for seafood is* fruits de mer — *fruit of the sea.*

Brittany, a grim, granite jawbone jutting into the Atlantic, is the great fishing province of France. Inhabited by a rude Celtic people, it was regarded in the 18th century, not as a romantic and picturesque vacationland, but as barbarous, backward, and poverty stricken. Its coasts were lined with the remnants of castles and huts of fishermen who divided their lives between using and fixing their nets. Their needles (3, 4), floats (9) and pins (5-8) are illustrated at the top along with two gauges (1,2) for measuring the mesh— or so the description says, for these devices smell not of fish but of the Encyclopedists' bent to geometrical edification. Neither are they in evidence at the right, where fishermen make nets (10, 11) from their balls of cord (13-15). On the wall at the bottom (19-22) is a crude cordwalk for spinning cordage, and a cauldron of tar for impregnating it.

Fishing I

Fig. 11. Fig. 17. Fig. 10.

Fig. 12.

Fig. 14.

Fig. 13. Fig. 15.

Fig. 16.

Fig. 24. Fig. 23. Fig. 22. Fig. 18.

Fig. 21.

Fig. 20.

Fig. 19.

Fig. 25.

Plate 43

Plate 43   Fishing II

The men fish at sea (fig. 1, q), the women from the shore, working their way out to improbable platforms from which they lower little hand nets. Heavier offshore nets were also used, of every imaginable description. That of Fig. 2 was known as a "sleeve". The boat is propelled by sculling, the universal seaside method, and it looks like a racing shell beside the ponderous rowboat of Fig. 3. When dragging for shrimp (Fig. 4), the boat is anchored fore and aft and the shrimp net handled like a great scoop.

Fig . 3 .

Fig . 2 .

Fig . 4 .

Plate 44

Plate 44   Fishing III

Oysters might be taken either by rakes (Fig. 1) or by dragging (Fig. 2). Unlike shrimp fishing, oyster nets were weighted down and dragged under sail. Oysters were also cultivated (Fig. 3) in artificial beds (b, e) prepared on the edge of salt marshes (a), after which they would be sown down the bay to build up the native population. Fig. 4, at the right, shows women at work, again. With their clothes piled on the bank, the wives of fishermen are seining the mouth to a small inlet. The artist leaves no illusions about what a life like this would cost in the way of female charms.

Fig. 3.

Fig. 4.

Plate 45

Plate 45   Fishing IV

The trident (Fig. 2), not yet a mythological instrument, was actually used in spearing fish. The shrimp net is being wielded again in the lower two vignettes (facing), from a small boat (Fig. 6) and wading in the shallows (Fig. 5). The construction of Fig. 4 is designed to trap tidal fish as the water ebbs and uncovers the beach.

Fig. 4.

Fig. 3.

Fig. 6.

Fig. 5.

Plate 46

Plate 46   Fishing V

The net in the water (Fig. 1, b) was known as a "sparrow-hawk." It is conical in form and weighted down with a lead sinker. The fisherman tosses it over a school of fish within reach. Strung along a cable (Fig. 2) are his eel-pots, some spherical, some bottle-shaped. So far these are the only fishing devices which required the use of bait. The nets of Figs. 3 & 4 are designed to trap larger fish, up to 5 or 10 pounds.

Fishing V

Fig. 3.

Fig. 4.

Plate 47

Plate 47   Fishing VI

Trawling and dragging nets.— The arrangement of Fig. 4 catches both surface and bottom fish.

Fishing VI

Fig. 3.

Fig. 4.

Plate 48

Plate. 48   Fishing VII

The tidal net of Fig. 1 has nothing unusual about it, but the fisherman of Fig. 2 employs a method which has fallen into disuse. He is seated in a blind, duck hunting with a pair of nets. His prey, the widgeon, passes down the coasts of France during the fall. It is a stupid bird and might also be entangled in the baited net of Fig. 3. To the modern conscience snaring birds by night (Fig. 4), the deadliest way of all, seems more treacherous than fishing by day. Attracted by the flare, the birds are blinded by it and can be engulfed in the net and dispatched at will. But for these people, sporting considerations did not enter the picture. They had to live on what they could catch.

Fishing VII

Vol. VIII, Pesches de mer, Pl. VIII.

Plate 49

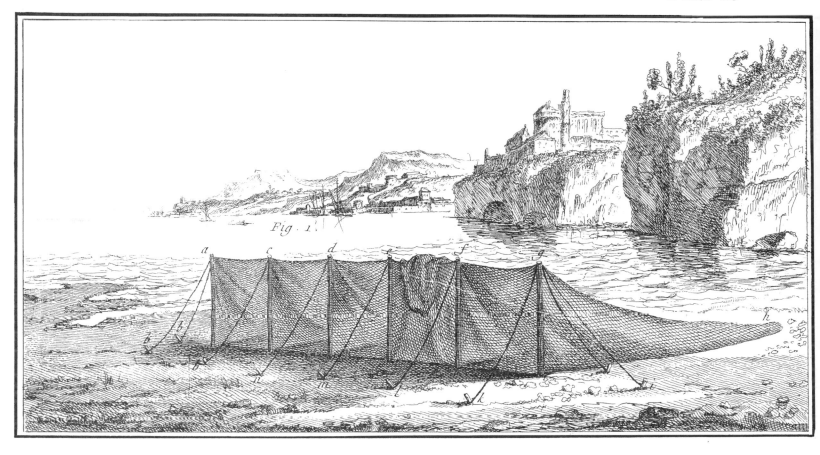

Plate 49   Fishing VIII

Herring will be the catch in Fig. 2, and mackerel of the coarser seine in Fig. 3.

Fig. 2.

Fig. 3.

Plate 50

Plate 50   Fishing IX

Some species of fish are more readily caught when they head into rivers for spawning. Mackerel are to be seined by the nets of Fig. 1, and up stream a bit the weir of Fig. 2 is equipped with salmon traps.

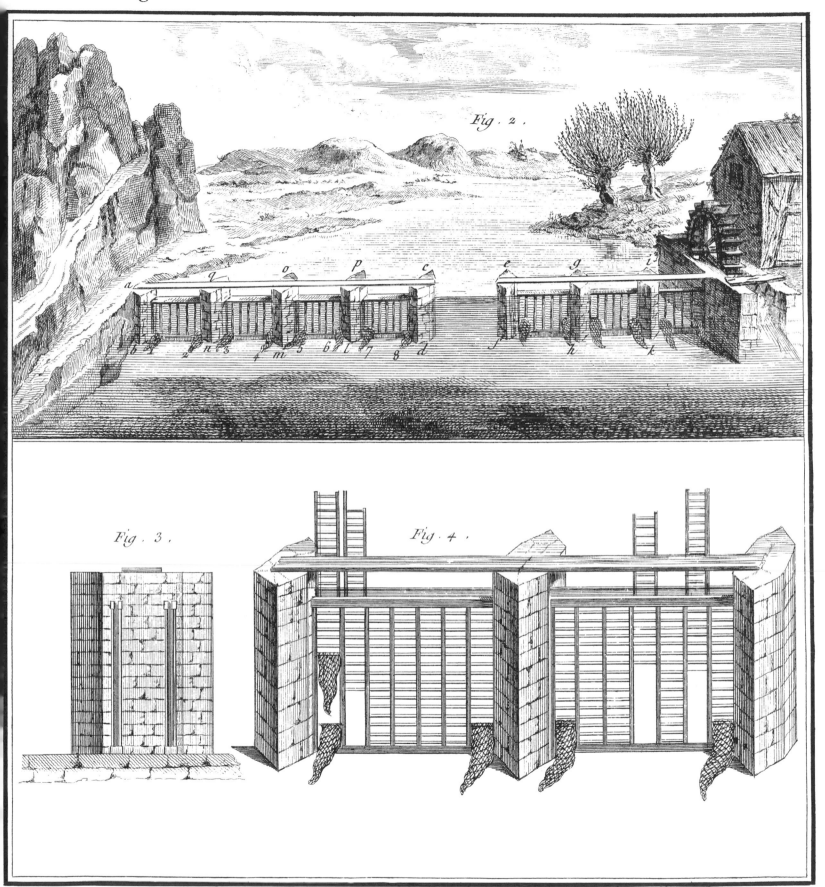

Plate 51

Plate 51   Fishing X

Drying, salting, and smoking were the chief means of preserving fish for the winter or for sale in distant markets. Sardines were the most valuable product of Brittany, which — as could scarcely be more evident — was then as now a poor province. The fish are brine-soaked in the combination warehouses and habitations, washed in the sea (Fig. 2), and packed in barrels (Fig. 3).

Plate 52

Fig . 1.

Plate 52   Fishing XI

Both sardines and herring were smoked (Fig. 1) in a most unsavory shed. In Fig. 2 is a tidal fish-weir and in Fig. 3 a net known in this arrangement as a "wolf."

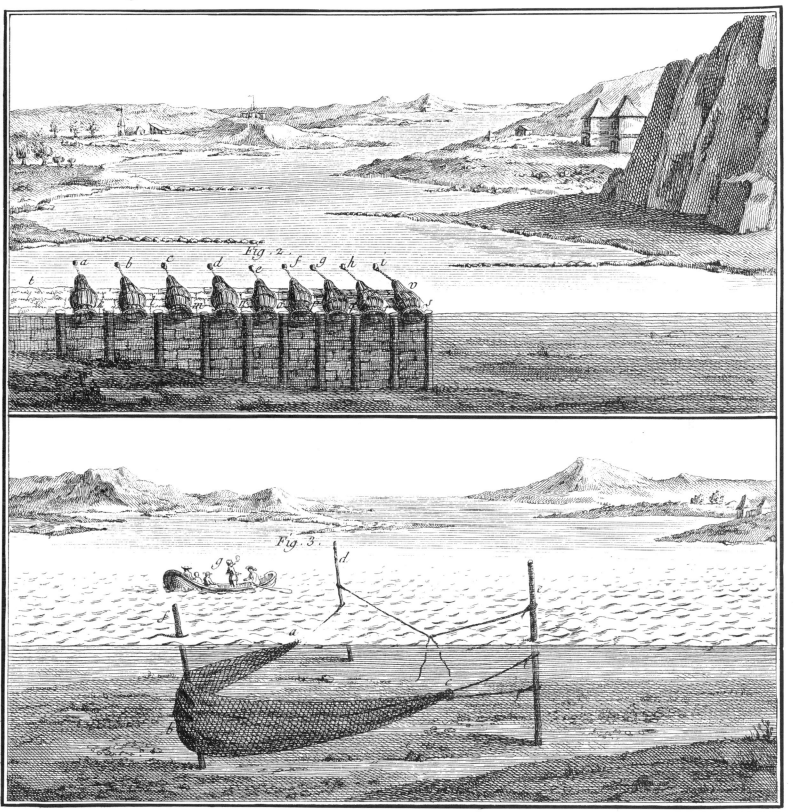

Plate 53

Fig. 2. *Fig. 1.*

Plate 53   Fishing XII

Fishing by flares at night was illegal but effective (Fig. 3). The fisherman of Fig. 1 wears pattens to keep from sinking into the muck.

Plate 54

Plate 54 Fishing XIII

*No fly fisherman this, for stream fishing too was a matter of food rather than sport.
The fisherman has provided himself with a portable fish screen to arrest the trout
where it can hardly fail to perceive one of the eight hooks proffered it, five on the
handline and three on the rod.*

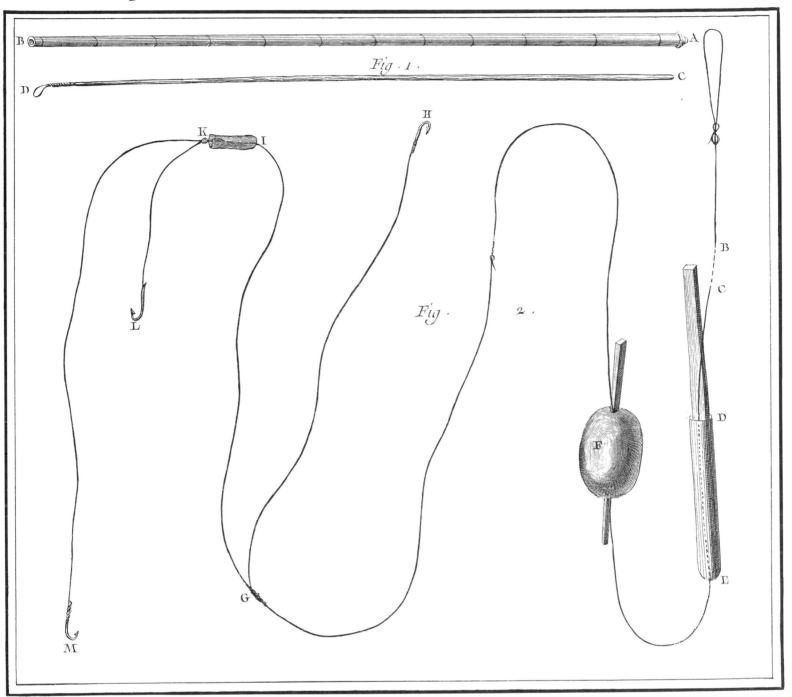

Fig. 1

Fig. 2

Plate 55

Plate 55   Fishing XIV

Netting trout or perch with a scoop net.

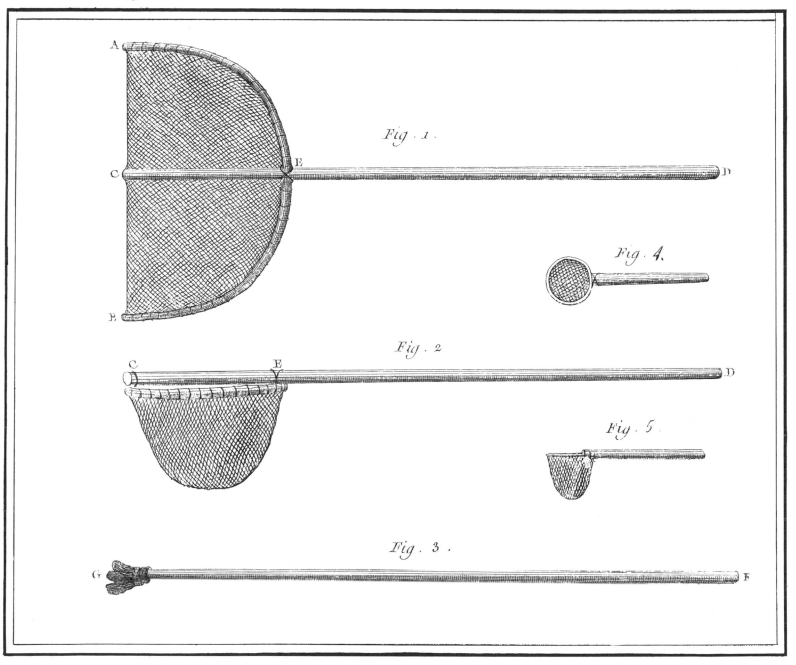

Fig . 1 .

Fig . 4 .

Fig . 2 .

Fig . 5 .

Fig . 3 .

Plate 56

Plate 56   Fishing XV

*In river trawling the net was cast upstream and handled by one man while his partner
kept the boat broadside to the current so as to maintain drag on the net.*

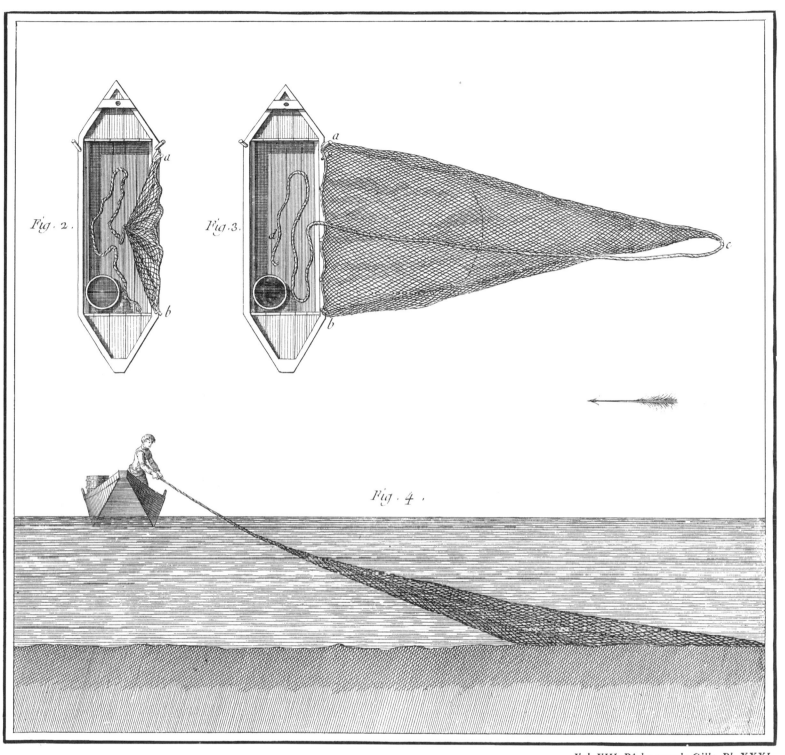

Fig. 2.

Fig. 3.

Fig. 4.

Art of War

Art of War

Historians are given to portraying the 18th century as an age of halcyon warfare. Not that wars were few. On the contrary, it is the classic age of dynastic conflict. But wars were relatively superficial: conducted not by whole peoples, but by mercenary armies; not raging across the whole countryside, but localized before particular strongpoints made for the purpose; fought not to the death, but to the point where one commander clearly could win. On this his opponent might decently throw in his hand like a poker player losing mainly treasure or a few possessions.

The picture must not be overdrawn, for men were killed in the battles of the 18th century. But it is true that the wars of that time, polarized around the Anglo-French rivalry, were political rather than ideological episodes. They were limited and restrained by civilized conventions as the wars that followed and preceded them were not. Wars of peoples, our own total wars, began only in 1792, with the struggle against Revolutionary and Napoleonic France. On the other hand, the wars of religion had been brought to an end by the Peace of Westphalia of 1648, and behind them there stretched the carnage of the Italian wars of the 16th century and the feudal conflicts of medieval chivalry.

The 18th century, then, represents a definite stage in the evolution of the military arts. Armies were not patriotic. But they had moved way beyond feudalism to become professional. This transformation was only the reflection in military affairs of profound changes in society. The chivalric knight had turned into a privileged gentleman. About all that his class had left of its warrior status and feudal authority over persons was a near monopoly of commissions as line officers. Beside the nobility and gentry there had grown up the bourgeoisie, and it was among them that officers were found for the new branches of military service which required technical skill, the corps of engineers and the artillery. Within the army itself, the professional soldier had replaced the feudal levy, and the foot soldier had moved into the position of primary importance once occupied by cavalry.

Technology had had a part in bringing about these changes. But it is possible that the revolutionary impact of the introduction of firearms has been exaggerated. The earliest cannon were unwieldy if portentous, and the effectiveness of massed infantry was first demonstrated, not by musketeers, but rather in the defeat of the mounted cavaliers of Charles the Bold, Duke of Burgundy, by the phalanx of Swiss pikemen in the battles of Morat and Nancy in 1476. By the 18th century the muzzleloading musket had reached the final stage of its evolution in the flintlock, which by attachment of the bayonet combined also the advantages of the pike. But cannon rather than small arms had the greater impact on the art of war — in an evolutionary sense, indeed, small arms themselves derive, not from archery, but from artillery. Cannon made the medieval castle untenable, and it was the superiority of the French artillery under Charles VIII which won his armies the run of Italy in the 1490's and opened the peninsula to the depradations of the more barbarous powers beyond the Alps.

The defeat of the Renaissance city-states did not spell the end of defensive fortification, however. On the contrary, it stimulated its improvement. Some of the greatest scientific minds of 16th-century Italy — Leonardo da Vinci, Niccolo Tartaglia, Galileo — addressed themselves to the perfection of military architecture. The art was carried back to France by the armies of Francis I (1515-1547), who lifted many aspects of Renaissance culture across the Alps to be acclimatized in France, and while medieval castles fell into ruin throughout the interior of the kingdom, royal fortresses of Italian inspiration grew up along the borders. From the late 16th century through the mid-18th, the development of military engineering and military organization centered in France and flourished under the stimulus of the dynastic warfare of the Bourbon kings. Modern military terminology still bears witness to this influence. No professional vocabulary is drawn so largely from the French as that of the army: it teems with words like reconnaissance and defilade, lieutenant and colonel, battalion and regiment, mortar and artillery, bomb and grenade — the list could be made very long.

The growing importance of the engineer culminated in the career of the famous Marshal de Vauban, who surrounded the France of Louis XIV with a chain of seventy-odd fortresses upon which he rested French security and based French power. It was behind this rampart that French culture flourished in its classic age and throughout its 18th century supremacy. Not that France's frontiers sealed her off from Europe — on the contrary, fortification served less as a barrier than as a military lightning rod. If Prussian or Austrian or English generals were busy with the ritual of siege warfare, why then the impact of conflict was harmlessly diverted from countrysides and cities and channeled to earth in devices made for war.

It was a convenient axiom that no strong point must be left untaken in a commander's rear. Wars were fought by just such formal rules. Their acceptance by both sides resulted in the oddly static quality which will be sensed in the plates that follow. War was, of course, a more damaging instrument of policy than diplomacy. But it, too, had its protocol. It, too, had its formalities. If they were observed, and if a defending commander was maneuvered into a position where he *should* be defeated, then surrender was not more dishonorable than is the act of a chess player who gives in when checkmate is inevitable. He would no doubt have lost some men in the course of his misfortunes. But he need not wait for them all to go. Nor need a defeat be irreparable. There can always be another game. There were, then, advantages in that heartlessness of 18th century international relations, which never took war too tragically, nor ever expected to see it abolished.

Plate 57

Plate 57 Medieval Men at Arms I

Two plates illustrate 15th century men-at-arms, whose equipment, tactics, and organization were to be rendered obsolete by the development of firearms and professional armies. As it still does, the art of war consisted of the application of fire and movement. Firepower was provided by the cross-bow, and movement and shock by cavalry. This horseman belongs to the light cavalry. He is armed with a sort of lance, but he might equally well be carrying a sword or even an arquebus. Both he and the ground soldier, a crossbowman, are protected by flexible coats of mail.

Various chivalric weapons are illustrated at right. They go all the way back to the bludgeons and battle axes (Figs. 15-18) which might have been used by Roland and Oliver at Roncesvalles, or even (Fig. 17) by the ancient Merovingian kings who pieced together the rudiments of the French state amongst the ruins of Roman Gaul.

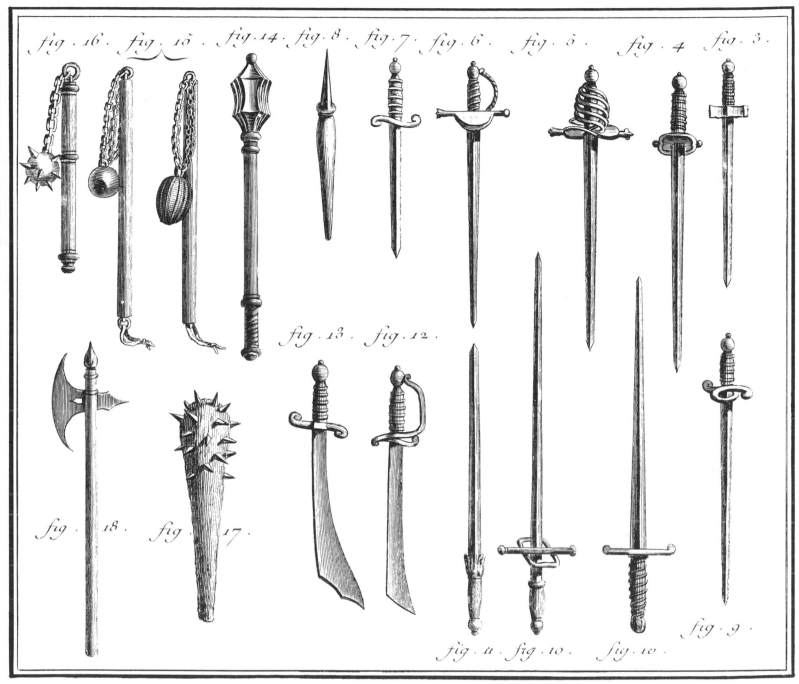

fig. 16. fig. 15. fig. 14. fig. 8. fig. 7. fig. 6. fig. 5. fig. 4. fig. 3.

fig. 13. fig. 12.

fig. 18. fig. 17.

fig. 11. fig. 10. fig. 10. fig. 9.

Plate 58

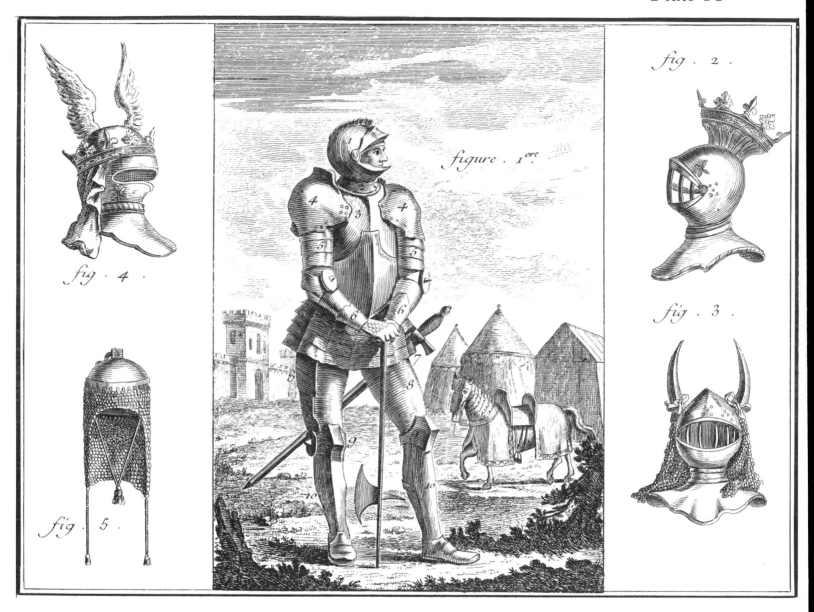

Plate 58 Medieval Men at Arms II

To the man at arms, afoot or on horse, corresponds the modern enlisted man, whereas the prototype of the officer was the belted knight. Here he is, with his mount and, in the background, his tents and some stronghold to be attacked or defended. Around the border are various items of equipment. Three helmets are shown (Figs. 2, 3, 4), each with a different crest to identify rank and person. The device of Fig. 2 is the royal insigne worn by the King of France and the princes of the blood. Under the helmet would be a bonnet of chain mail (Fig. 5). The horse too, is protected by a casque (Fig. 6). Men-at-arms carry shields and bucklers (Figs. 7-10). Those of knights would bear an appropriate coat of arms (Fig. 11). Bowmen attacking a stronghold would carry a shield (Fig. 12) large enough to shelter the whole body, and from be-

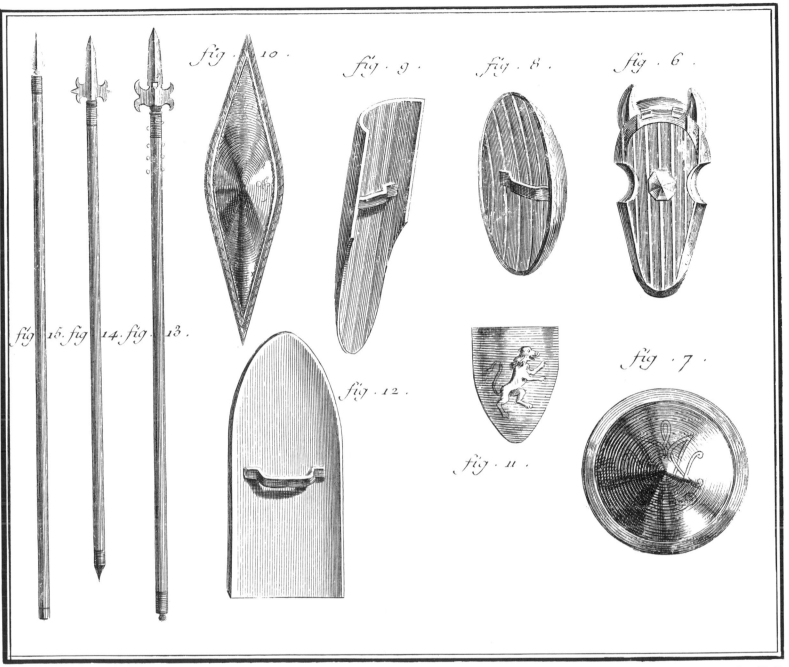

hind it they would launch their bolts and manipulate their clumsy crossbows or, in later years, muskets. Lances and pikes appear at the left, but the varieties used are too many to illustrate and the Encyclopedist concludes with Father Daniel, author of La Milice française, "that besides the sword and the lance, the knights and esquires used as weapons all sorts of instruments, provided only that they were suited to beating an enemy to death."

Plate 59 Small Arms I

The workshop, and all it stands for, was the enemy that the knight could by no weapon overcome. Here is a machine for boring musket barrels, side-view (Fig. 1) and top-view (Fig. 2). The power is external, and the working of the mechanical linkage by which it is communicated from the big pulley-wheel (D) to the four bits (P) is self-evident. The water serves, not to drive the machinery, as might be supposed at first glance, but to cool the barrel during boring. The borer at the left shows a barrel clamped into place (S, T). Figs. 3 and 7 represent forges — they appear to be identical and the artist must have decided to give us two for symmetry. In front of each is an anvil — one double (Fig. 4), one single (Fig. 6).

Plate 59

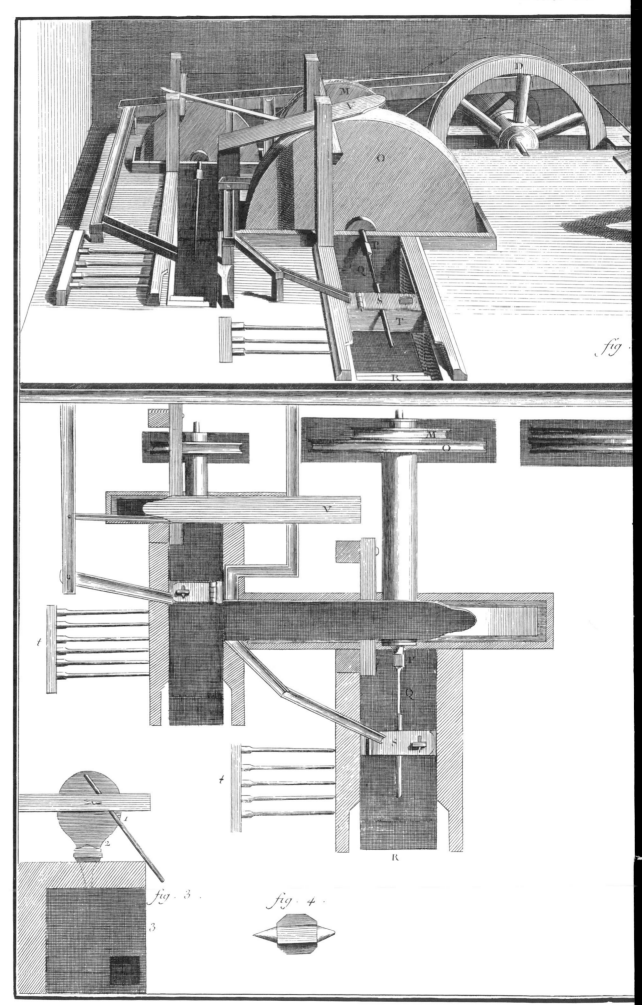

fig.

fig. 3.

fig. 4.

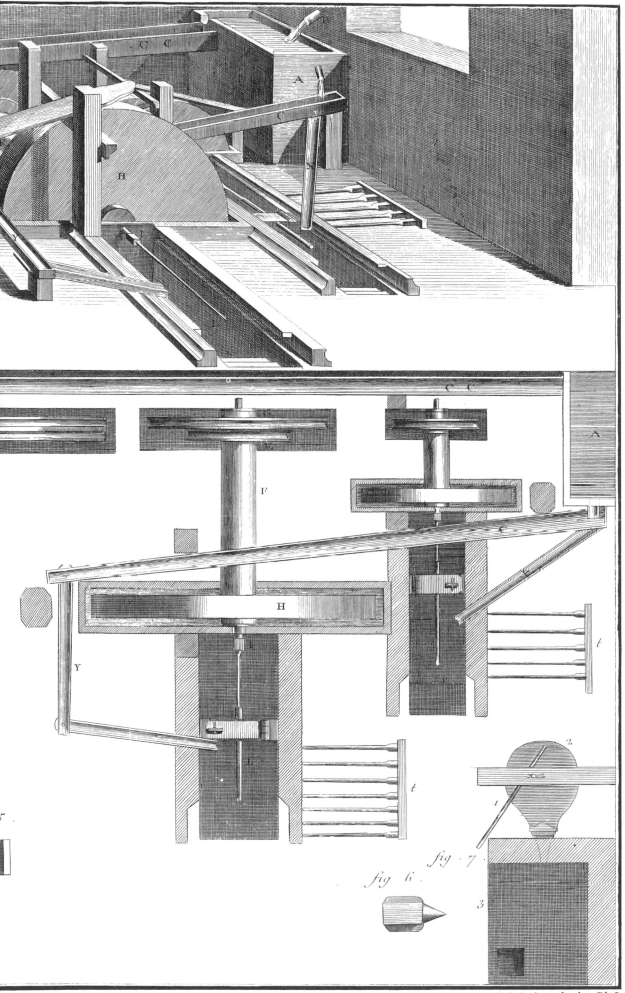

fig · 7 ·

fig · 6 ·

Plate 60   Small Arms II

This is a machine for grinding and polishing or decorating the exterior of the barrel. Few European muskets were rifled in the 18th century. It is, indeed, unclear whether rifling was introduced to cut down fouling or to produce true flight. For the projectile was still a ball, and only the modern ogival bullet will draw the full measure of accuracy from spinning.

Plate 60   Small Arms II

Plate 61   Small Arms III

The flintlock was introduced into the French army in the first years of the 18th century. The problem around which the design of small arms revolved, and which produced their evolution, was that of striking fire to the powder. In the earliest muskets a red-hot wire was touched to the powder through a touch-hole. This was followed by a glowing wick, later attached to the piece itself in the 15th century matchlock. The next step was the 16th century wheel-lock, igniting the powder by striking sparks from a steel wheel released from a clock-spring tension to spin against a fragment of pyrites attached to a cock in contact with the powder. Finally, in the snaphaunce musket the mineral element — a piece of flint (Fig. 1, D) — is transferred to the hammer (Fig. 1, B, C) and in the final refinement of the flintlock the pan (i, G) and stud (i, K) become a single piece (Fig. 4).

Fig. 2 shows the cocking and firing mechanism in detail, Figs. 5 and 6 the flintlock of the infantry soldier, and Fig. 7 the flintlock pistol, carried by officers and cavalry. The weapon fired a round ball loaded by the muzzle — see the ramrod (Fig. 6). Breech-loading and the use of elongated and ogival bullets fired by a percussion cap were the 19th century steps in the evolution of the infantry weapon, and the introduction of semi-automatic and automatic fire are contributions of our own time.

Plate 61   Small Arms III

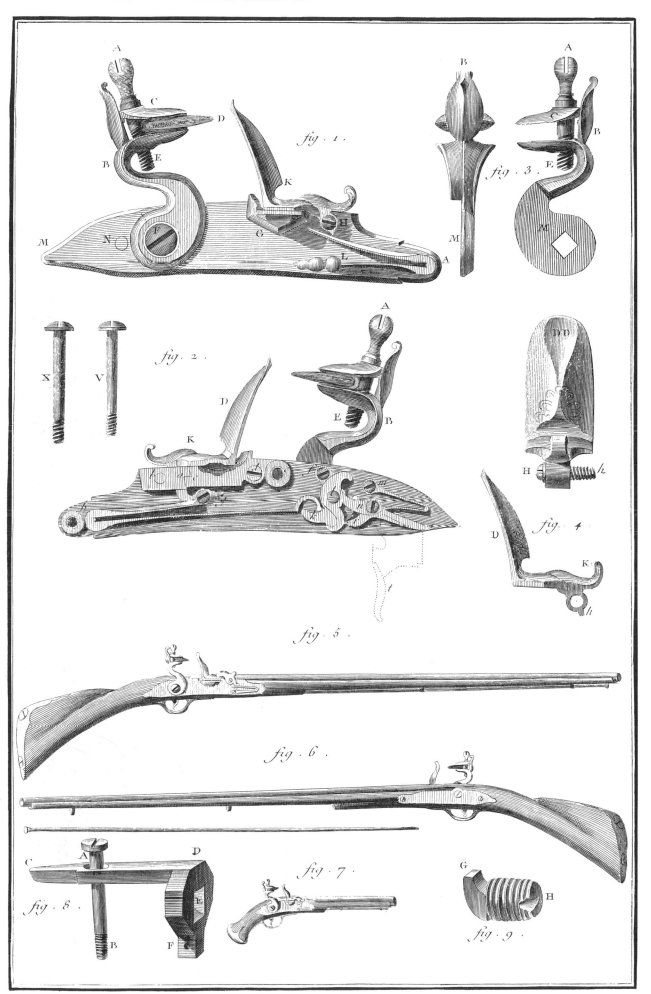

Plates 62, 63, 64, 65, 66   Drill & Maneuver I-V

Today's draftee in basic training resents not only the drudgery but even more what he considers the empty formalism of close order drill and the manual at arms. And indeed these relics of the military practice of the old regime, greatly simplified though they have been, do serve only to instill discipline and regimentation into a body of men. No one fights by the manual of arms. In the 18th century it was otherwise. Drill was at once an instrument of discipline and a training for combat, in which men fought shoulder to shoulder in a ritual as regulated and ordered as a minuet, and as purposeful. Inspection of the positions illustrated in the plates that follow will show two things: the relative poverty of the modern manual of arms, and the fact that each posture is related to the use of the weapon: for example, (1-9), fixing the bayonets for the charge; 25-29, Ready, Aim, Fire; (30-36), Reload and cock — notice the cartridge to be torn open by the teeth (33). Try running your eye quickly along the series as if it were a sequence of movie stills — you will have a sense of the motion of the drill.

Plate 62 Drill & Maneuver I

Plate 63   Drill & Maneuver II

Plate 64   Drill & Maneuver III

Plate 65   Drill & Maneuver IV

Plate 66   Drill & Maneuver V

Plate 67

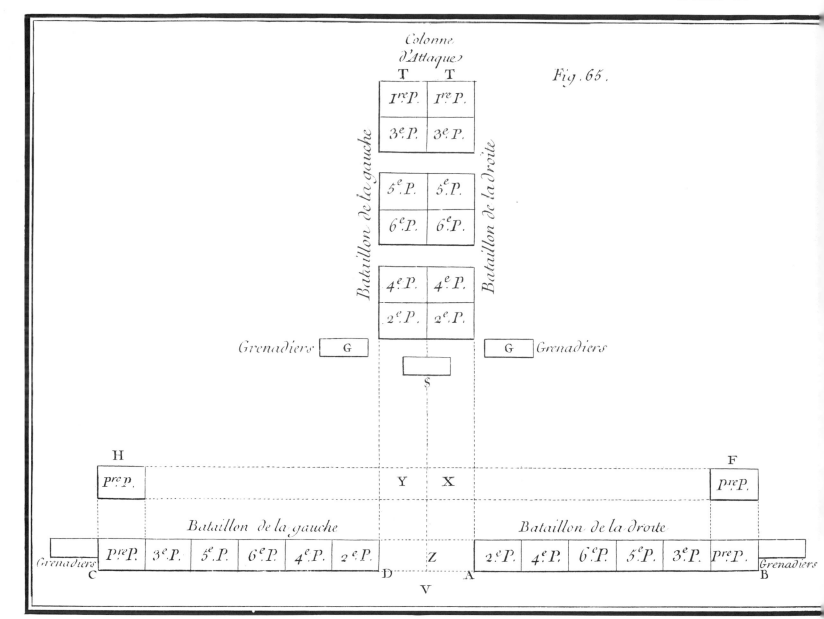

Plate 67   Drill & Maneuver VI

This may serve as an illustration that the order of battle of the 18th century survives as the order of parade of a modern army. If this were a modern regiment it would be about to pass harmlessly in review before its commanding officer or some visiting dignitary. It is, however, an 18th century regiment composed of two battalions which have just formed from line into column of attack. A diagram of the maneuver appears above (Fig. 65). AB and CD are two battalions, of six platoons each, abreast in line formation. In advance of each flank are the leading platoons (H, F), serving as "points." These platoons march to the center (X, Y) and the column of attack forms behind them in three elements, with the grenadiers at the rear on either flank (G). At the rear (S) marches a platoon of supernumeraries.

Fig 66

3.ᵉ Section · 2.ᵉ Section · 1.ʳᵉ Section

Plate 68

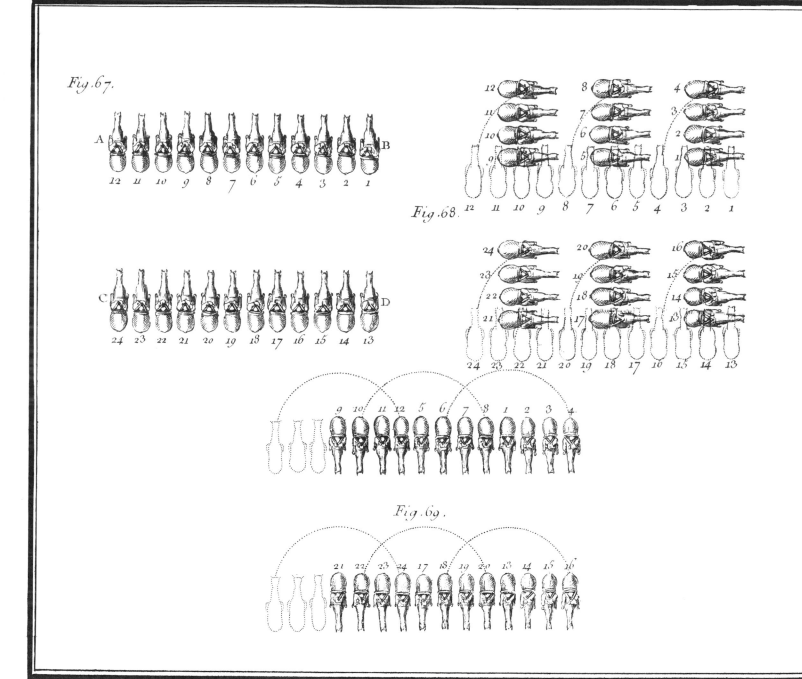

Fig. 67.

Fig. 68.

Fig. 69.

Plate 68   Drill & Maneuver VII

This plate illustrates cavalry maneuvers. Animals are shown from the top in Figs. 67, 68, and 69 to make the movements clearer. The two lines of cavalry in Figure 67 are to be disposed into columns of four. This is done by a maneuver corresponding to the old "Right by squads, march." The evolution to the rear is done by a further quarter turn. Fig. 70 illustrates formation of marching columns from a 168

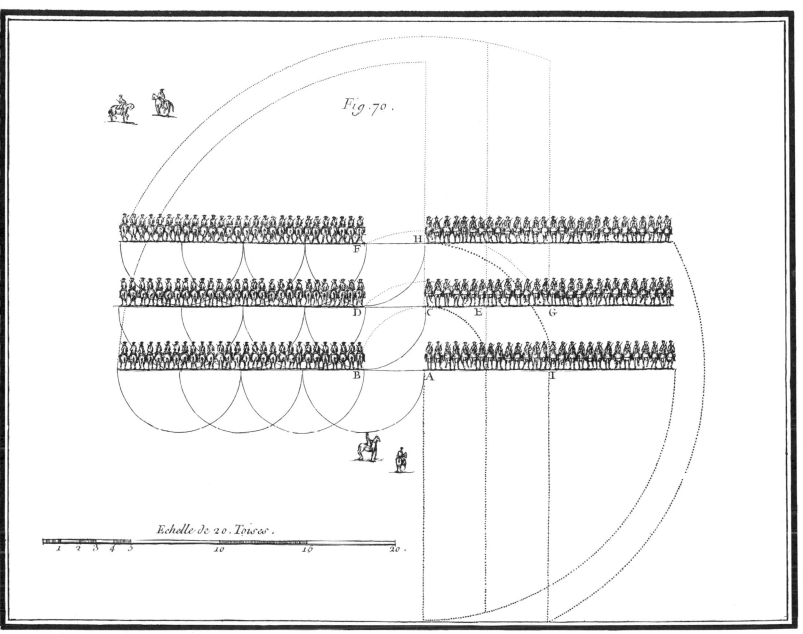

man squadron composed of four companies of 42 cavalrymen. The commanding offi-cers supervise the complicated maneuver. The scale at the bottom indicates how close close-order was in the 18th century. Horses, like soldiers in the infantry, literally stood shoulder to shoulder.

Plate 69

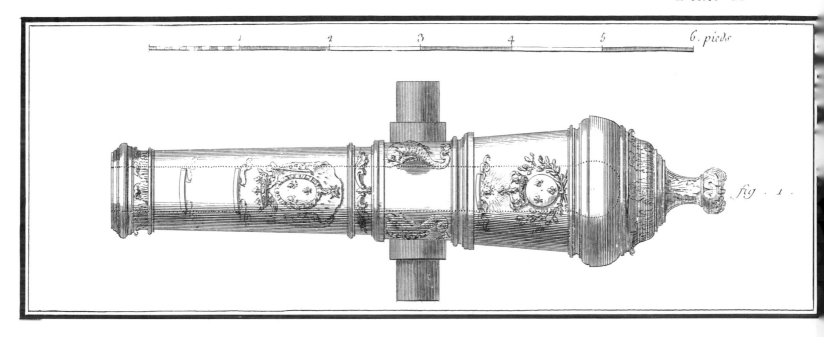

Plate 69   Artillery I

It was upon his artillery that an 18th century commander relied for breaching the fortified places around which warfare revolved. Ordnance is illustrated in the following plates. This is a 24-pound bronze cannon — i.e., it fires a 24 pound shot. It is almost eight feet long. The design of bore and chamber are indicated in dotted outline (Fig. 1). Notice the embellishment. The earliest cannon were cast by men who worked in the atmosphere of Renaissance craftsmanship, and the tradition persisted that an artillery piece should be, not only an instrument of destruction, but a work of art, worthy of its maker and its user. (For the technique of casting and boring cannon, see Plate 108ff.).

In Fig. 2 the gun is mounted on its carriage, details of which are evident in Figs. 3 and 4, and in Fig. 5 it is attached to the fore-carriage by which the piece is transported in the artillery train and wheeled into position. Veterans of the first World War will remember horse-drawn artillery pieces that had to be pulled backwards into place and swung around to face the enemy. At the right (Fig. 6) are tools that serve the piece, and at the bottom (Fig. 7) the parts of the gunsight.

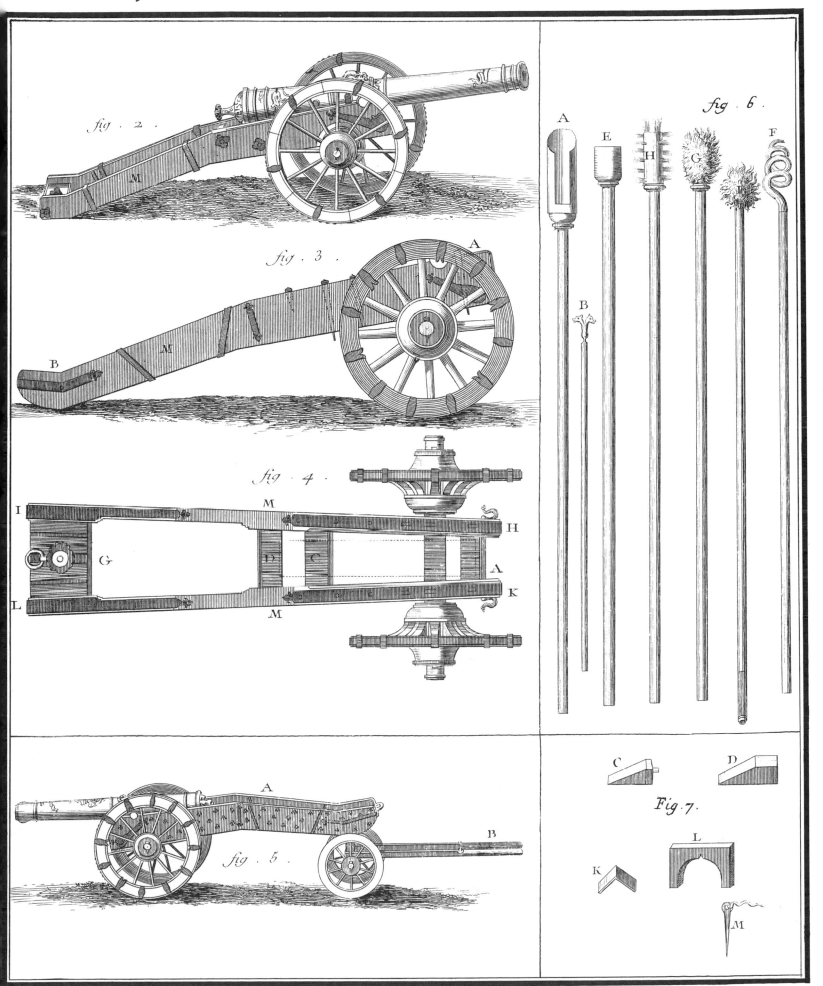

fig . 2 .

fig . 3 .

fig . 4 .

fig . 5 .

fig . 6 .

Fig . 7 .

Plate 70 Artillery II

The cannon was a flat-trajectory weapon. Fig. 1 illustrates how the ball would deviate from the line of sight. This depended on the weight and characteristics of the ball, and gunners used a compass (Fig. 3) for calibrating the ammunition.

The mortar (Fig. 4) was the weapon of high-angle fire. It had two advantages. It could reach a target in defilade — i.e. hidden behind a hill or rampart — and since it lobbed its shell at very low velocity, it could fire an explosive charge, a bomb (Figs. 5, 6, 7) rather than the solid cannonball. The range of the mortar depended on the angle of elevation of the barrel, and this was adjusted by a quadrant (Fig. 8), details of which are given in Fig. 9. This device was identical in principle with the sight used in the early part of World War II to lay the 4.2 inch mortar of the U. S. Army Chemical Corps.

Plate 70   Artillery II

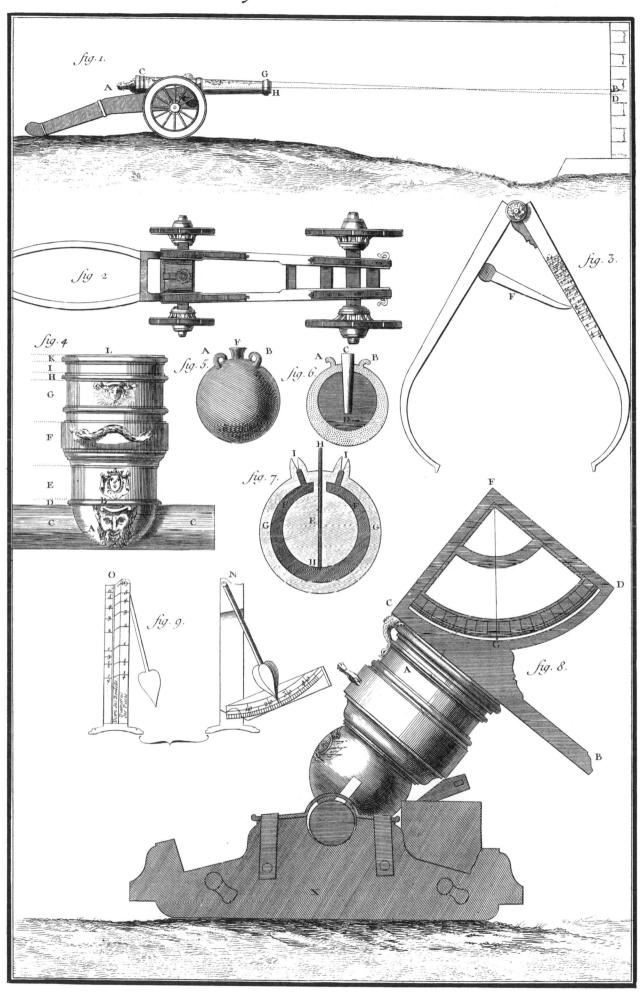

Plate 71 Artillery III

Fig. 1 illustrates the trajectory of a mortar shell, and Fig. 2 the geometry concerned. The mortar in Fig. 3 was designed to fire rocks — a cheaper form of shell and one less given to blowing up in the mortar (exterminating the guncrew instead of their enemies) rather than at the target. This remains today a possibility which no experienced mortar-man dismisses from his mind. The shell was a framework of steel bands covered with canvas and filled with stones (Fig. 4, 5, 6).

Fig. 7 is an early howitzer, a form of artillery which represents a cross between the cannon and the mortar. Beside it is a petard (Fig. 8), and a view of the under side of the plank on which it was mounted (Fig. 9). Petards contained a charge of powder and a fuse and were used to blow in a wall or a fortified gateway.

Finally, in Fig. 10, there is a top-view of an artillery battery emplaced. The commander has had a ditch dug out (fossé) to protect his front and flanks. His pieces are set on a foundation of planks. Number One gun is omitted to show the construction (B, C). The battery is protected by a rampart (D, F), in which embrasures (E) provide a field of fire. From the scale at the bottom, it is apparent that the guns are placed almost fifteen feet from muzzle to muzzle.

Plate 71 Artillery III

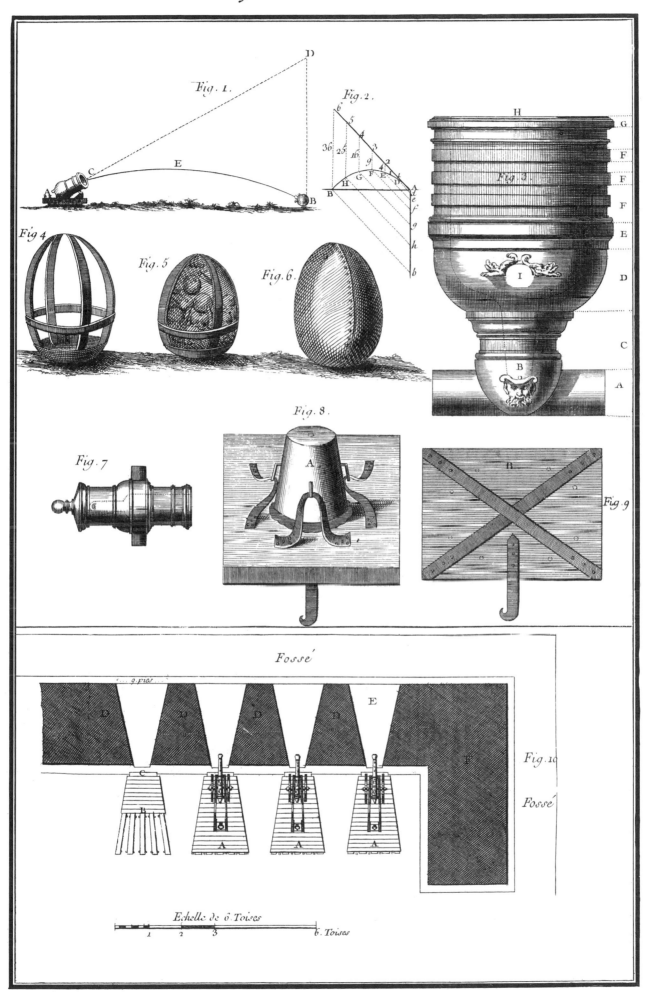

Fig. 1.

Fig. 2.

Fig. 3.

Fig. 4.

Fig. 5.

Fig. 6.

Fig. 7.

Fig. 8.

Fig. 9.

Fig. 10.

Fossé

Fossé

Echelle de 6 Toises

6. Toises

Plate 72   The Art of War I

Fig. 1 is a profile of the gun emplacement just illustrated, and Fig. 2 a top-view of a mortar battery in place. D is the parapet over which the mortars fire, and the construction of the platform is illustrated at B, C. It must be very sturdy, for while the cannon recoils backward, the mortar tends to drive itself into the ground. Fig. 3 is a profile of a mortar in the act of being fired. Notice the aiming stakes (f, g) on the parapet, which will look familiar to World War II heavy mortarmen — except that as the accuracy of the weapon increased, it became necessary to place them farther apart and farther out.

Plate 72 The Art of War I

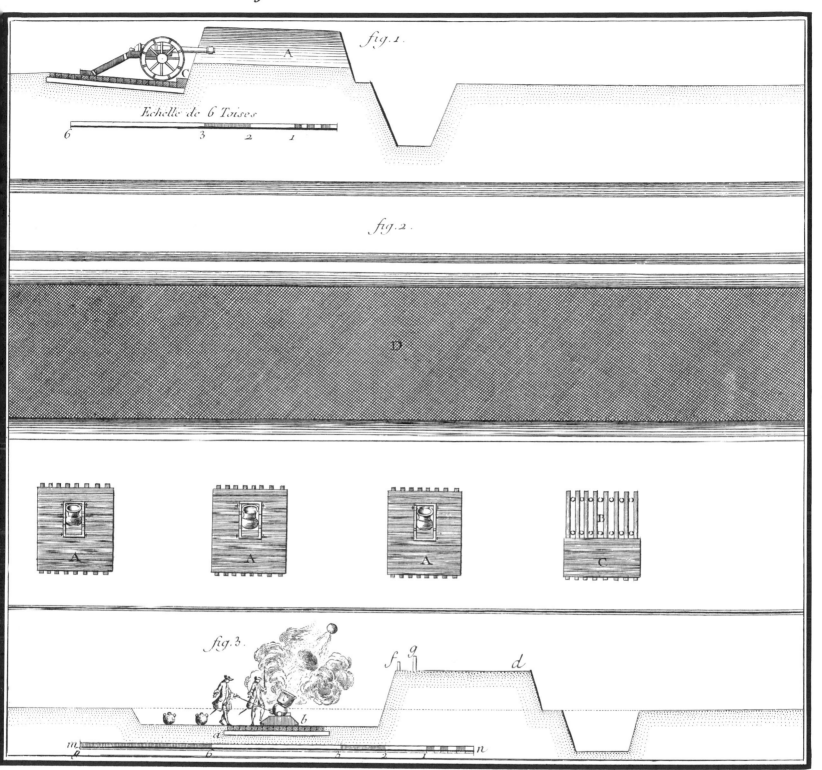

fig.1

Echelle de 6 Toises

6   3   2   1

fig.2

D

fig.3

Plate 73

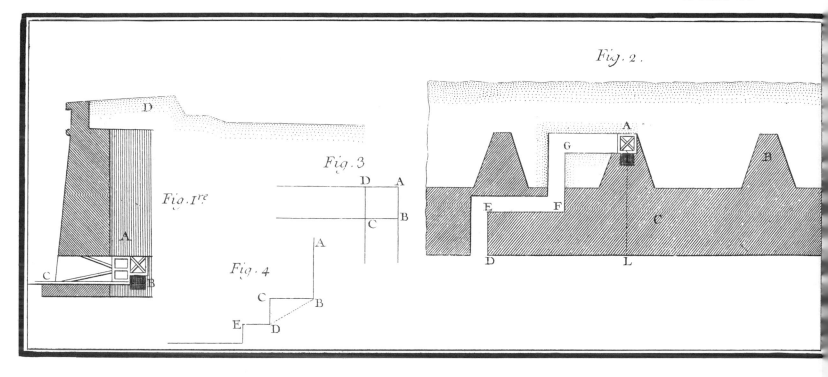

Plate 73   The Art of War II

This plate illustrates the employment of the sapper's weapon, the land mine. Fig. Ire shows a mine emplaced beneath the buttress (A) of a rampart under which the sappers have tunneled a passage from the outside C. The point is that if the distance B C is less than the height of the wall B D, the effect of the explosion will be partially wasted since it will be directed back along the passageway rather than upward into the body of the wall. Fig. 2 is a horizontal cross-section of the base of a fortification into which a zig-zag passage (E, D, F, G, A)(D, E, F, G, A) has been wormed in order to plant a mine (L). The advantage is that all these turnings bottle up the force of the explosion to give it maximum effect.

Fig. 5 is a double mine emplacement (AA) and Fig. 6 a triple one (AAA), each with its powder train (D, B, A). Fig. 7 is a profile view of the mine chamber (a), which is to be filled with explosive on a foundation of straw and sandbags. At the right (c) is the construction which the sappers erect as they dig out the earth in order to support the wall they mean to blow up. They light the powder train at (d) and then scramble back out of the tunnel they have dug. Fig. 8 is a front elevation and Fig. 9 a cross-section of the rampart to be destroyed — and Figs. 10 and 11 the same views a moment later with the mine "in play." And, finally, Fig. 12 shows all that is left after a good demolition. The wall is breached, and it is the turn of the infantry to work its way across the rubble — see Plate 81.

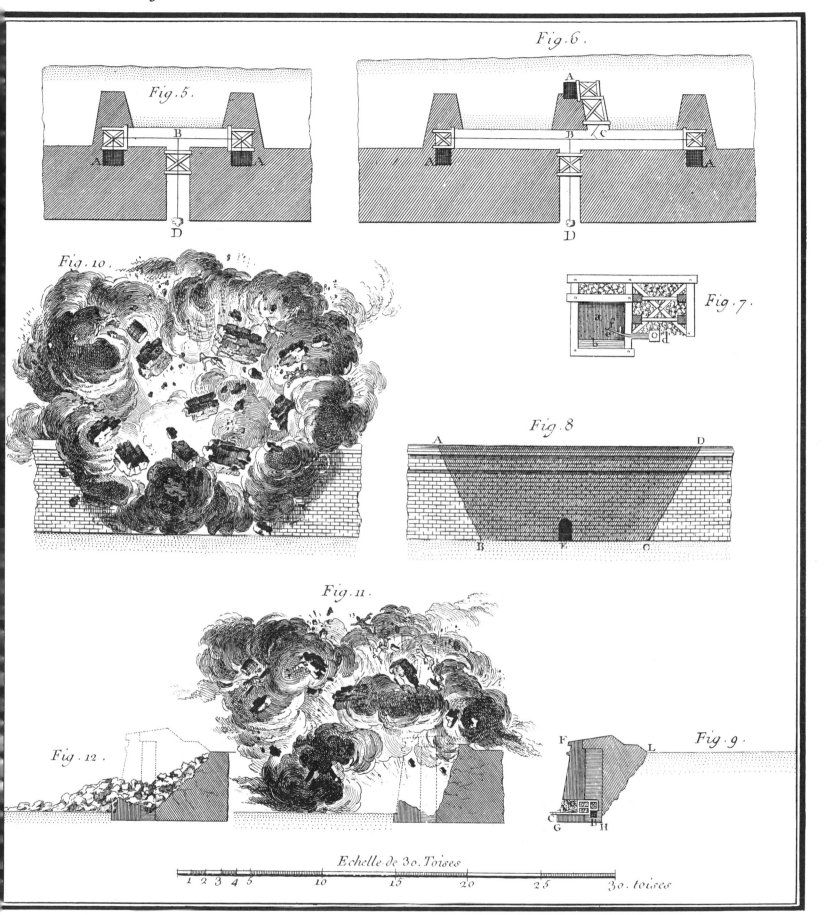

Fig. 6.

Fig. 5.

Fig. 10.

Fig. 7.

Fig. 8.

Fig. 11.

Fig. 9.

Fig. 12.

Echelle de 30. Toises

1 2 3 4 5   10   15   20   25   30. toises

Plate 74   The Art of War III

Here are various oddities. Fig. Iere is an obsolete assault tower — notice that "battering ram" is a metaphor taken literally by the military engineer of early times. Figs. 2, 3, and 4 are proving devices for powder. The pistol-shaped affair of Fig. 2 tests the strength of the powder by the distance that its explosion turns the ratchet-wheel (H) against a spring (I). The principle of 3 and 4 is similar. Fig. 5 is a test-mortar.

But it is Fig. 6 which is the real curiosity. This is the floating mine or "infernal machine" with which the English attempted an assault on Saint-Malo during the wars of Louis XIV. The apparent scale is deceptive, for this was a vast engine of destruction. In the bottom (B) was sand, for ballast. Then came a layer of powder-barrels (20,000 pounds of powder) another (D) of 600 shells of the rock-thrower type (see Plate 71), above that (E) fifty iron-bound casks "filled with all sorts of explosives," and finally a frosting, so to say, of old gun barrels and other pieces of scrap.

In operation, however, this "infernal machine" created "more noise than damage" and blew up harmlessly outside the great ramparts that look across the channel toward England. In 1944 the Germans enjoyed greater success, indeed almost total success, with the destruction of Saint-Malo, not in attacking the city, but in the act of surrendering it.

Plate 74   The Art of War III

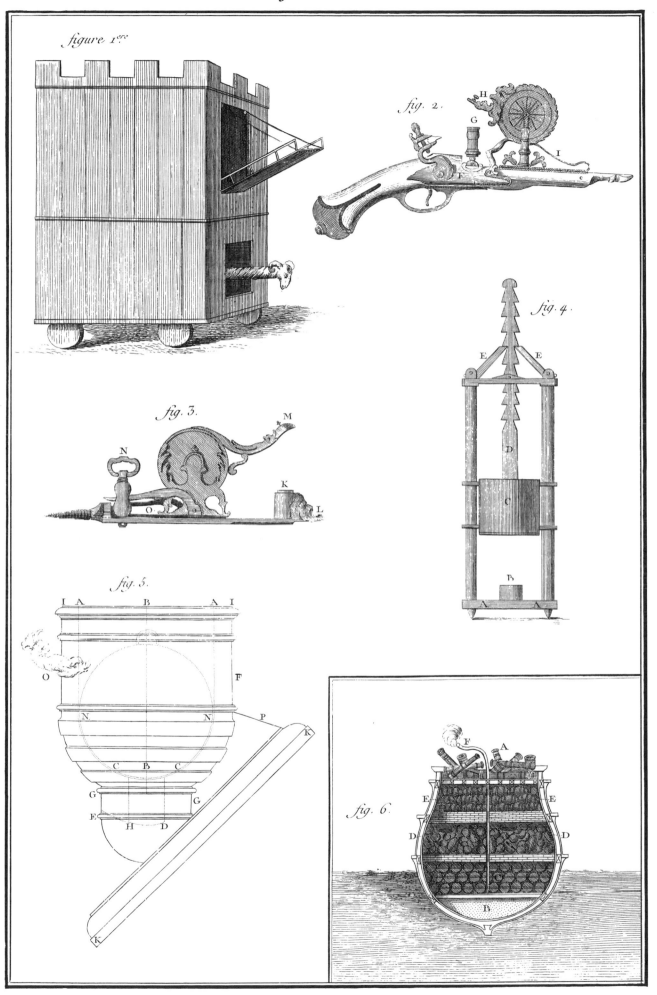

figure 1.ere

fig. 2.

fig. 3.

fig. 4.

fig. 5.

fig. 6.

Plate 75   The Art of War IV

Fig. I illustrates in relief the typical fortress which it was the main business of an 18th century commander to defend or to attack. Its nucleus was a pentagonal figure, or trace, outlined by a rampart with parapets. This was the main line of defence. A bastion (A, B, C, D, E,) was built out from each corner. These bastions were constructed so that the defenders could keep an enemy under cross fire, no matter from what direction he approached. In a well-constructed fortress there was no point in the surrounding area which could not be fired upon by two bastions, and bastions had to be within musket shot of each other.

Around the trace were the outer defences. A ditch, shown in dots, was bounded on its outer edge by the contrescarpe (a, b, d), an earthen bulwark. Behind the parapet of the contrescarpe ran a "covered way" which was connected by passages (g) with the inner fortress. Finally, the outer front of the earthworks, the glacis (g), slopes down to the terrain.

The remaining figures illustrate the principles of fortress architecture: the square and round towers of the parapet (Fig. 2); the "dead-space" beneath the parapet (3); interdiction of the area between bastions (4); and the detailed design of ditch and glacis (5). Fig. 6 shows how not to design the contrescarpe (HI). If it is parallel to the line of the trace (B C), then the flank (A B) of one bastion does not command the face (D E) of the next. Fig. 7 makes clear the advantage of bastions with concave flanks (D, G, H, I), and Figs. 8 to 10 illustrate the latest (8) and the older, less elaborate design of the "curtain" — the area between two bastions.

As admirers of Tristram Shandy's Uncle Toby *will remember, the exact angle, pitch, and placement of the elements of a fortress are very nice matters — involving infinite art and not a little pedantry.*

Plate 75 The Art of War IV

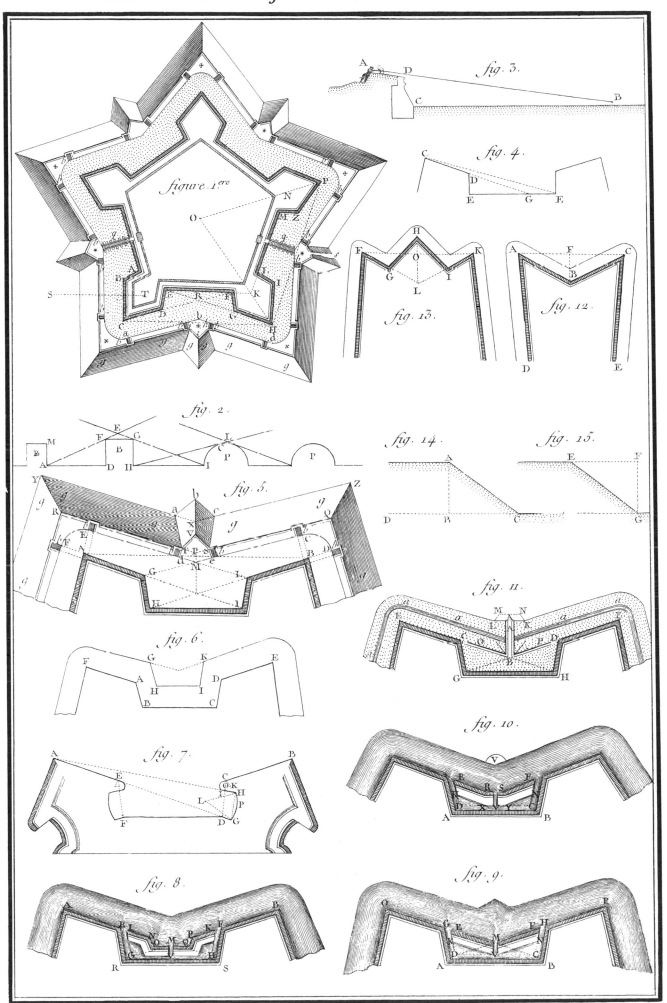

figure 1ere

fig. 3.

fig. 4.

fig. 13.

fig. 12.

fig. 2.

fig. 14.

fig. 15.

fig. 5.

fig. 11.

fig. 6.

fig. 10.

fig. 7.

fig. 8.

fig. 9.

Plate 76   The Art of War V

The 18th century fortress evolved by gradual stages out of the principles devised by Italian military engineers of the 16th century in their vain attempt to stem the devastating onslaughts from France, Spain and Germany. Fig. 1 represents the system of the first of the great French military engineers, Errard de Bar-le-Duc, who served Henry IV (1589-1610). Fig. 2 is the Dutch system, with which the House of Orange defended the Low Countries against Louis XIV (1643-1715). The remaining designs lead up to the work of the most famous of all military engineers, the Marshal de Vauban, who organized the defenses of France during the wars of Louis XIV. Vauban was more than a general. His career was the military expression of the genius of the Enlightenment. Through him speaks the rational and geometrical spirit of Descartes, bringing science and reason to bear upon warfare to better the condition of those who must engage in it. Both on the defense and the offense the tactics of Vauban were devised to make maximum use of engineering imagination and topographical analysis in order to save lives.

The line of fortress evolution lay in creating greater depth for defensive fields of fire. In Figs. 1-5 this objective is sought by extruding the bastions farther and farther from the trace of the fortress, and giving them more and more intricate flanks. Fig. 7, Vauban's first system, represents no essential departure. The improvements are details: powder magazines (I) in the bastions themselves; ramps (MN) to allow the cannon to be wheeled into place; a sounder parapet (BS) which runs all around the fortress, and even higher defenses in the bastion (L).

It is Vauban's second system (Fig. 8) which introduces novelties. The most important change is a separation of the bastions from the primary enceinte where they are replaced by smaller towers (BKL). The bastion itself is thrust out into the ditch. Advantages are twofold — an increase in that depth of fire for which the bastions had been straining outwards ever since Errard, and a more flexible defense. The loss of a bastion no longer means a breach in the main fortress. Vauban's third design (Fig. 9) further elaborates these principles. It served as the plan of the construction which crowned his career, the fortress of Neuf-Brisach.

Plate 76 The Art of War V

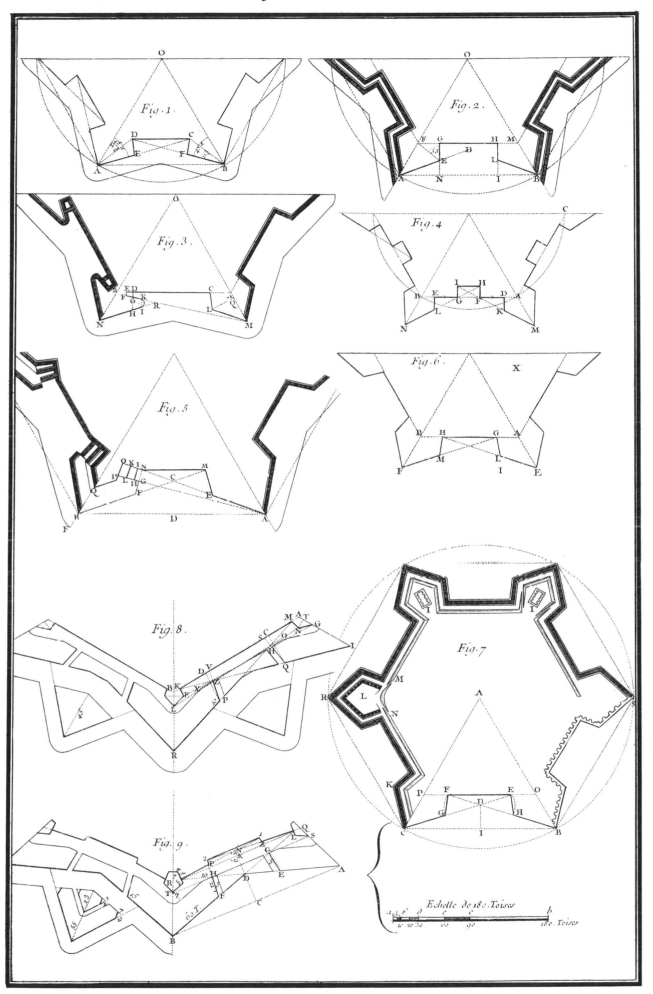

Plate 77   The Art of War VI

This plate combines details of fortress architecture with a relief view of a whole fortress built in Vauban's later manner. Fig. 1 is a cross-section of the main elements of a fort. In the center is the ditch, to the left is the main rampart (AE) behind the parapet (LMN) surmounting the escarp (RN). To the right outside the ditch is the contrescarpe (u, m) with its revetment (zy), above which the covered way (mc) runs behind a shallow parapet (ch). Finally, the glacis slopes gently down to ground-level (AB).

Fig. 2 gives plans for the detached "half-moons" (5) and "counterposts" (7) of Vauban's final system, while Fig. 3 illustrates these points in relief. Figs. 4 and 5 are variations on the theme of detached, mutually supporting strongpoints. Finally, Fig. 6 shows the ensemble of a citadel. Notice the dimensions: the radius (CB) of the enciente measures about a quarter of a mile. Behind the fort lies the esplanade (XY) of the city to be defended. The fort bristles out toward the open country with all the symmetrical intricacy of which the art of fortification was capable.

It was sculptural art, and it answered to certain deep formal instincts of French classicism. Compare the conception of these structures to the design of the formal gardens of this same reign of Louis XIV. There is the same elegant geometry of balance and proportion. LeNôtre, the landscape gardener of the king, and Vauban, a Marshal of France, thought in the same way. This style is equally apparent in architecture, where its instinct for symmetry still gives the French city its distinctive quality. This is particularly noticeable in the frontier cities of France, even though most of them have grown out beyond the ring of Vauban's walls. These plates make it obvious that the influence of French taste on Italian exuberance had the same effect in military as in civil architecture. It curbed imagination by geometry and fancy by reason.

Plate 77   The Art of War VI

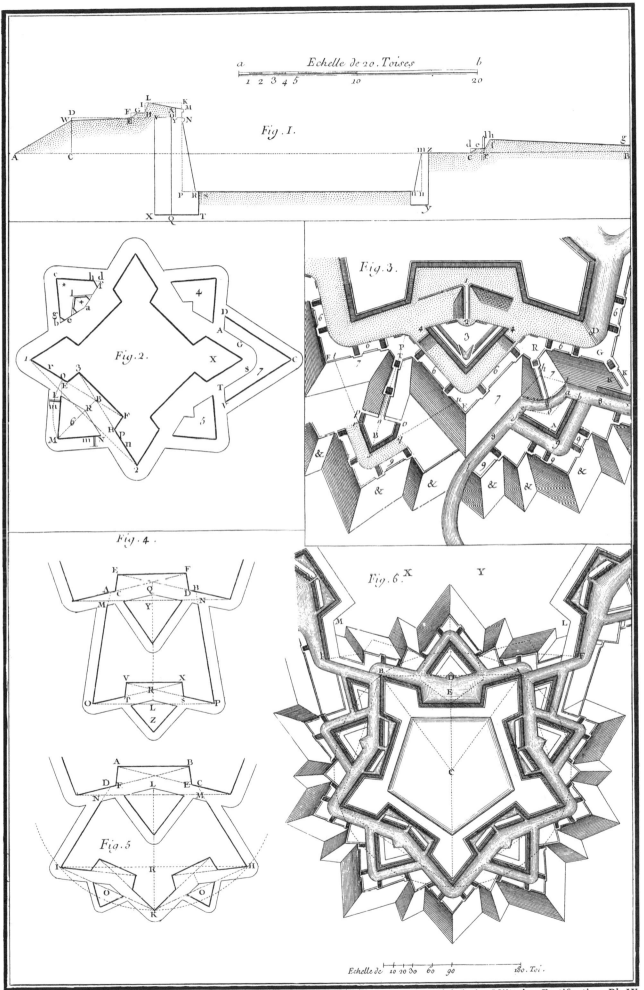

Echelle de 20. Toises

1 2 3 4 5   10   20

Fig. 1.

Fig. 2.

Fig. 3.

Fig. 4.

Fig. 5

Fig. 6.

Echelle de 10 20 30 60 90   180. Toi.

Plate 78

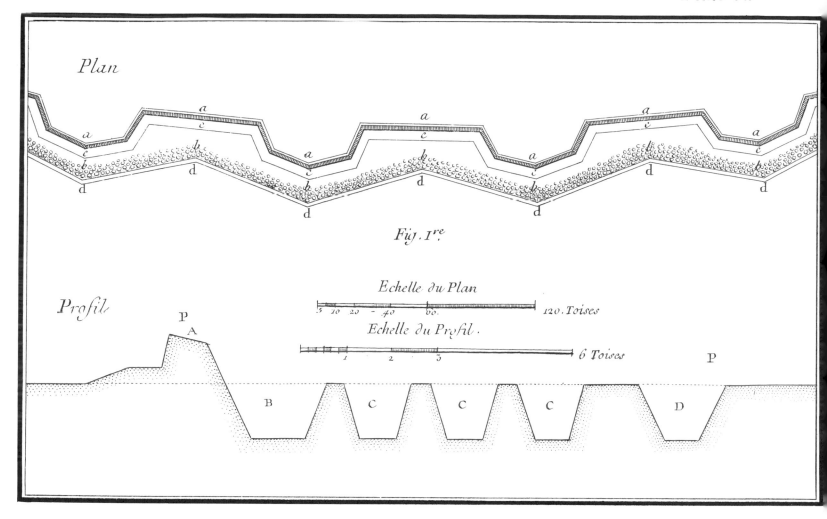

Plan

Fig. Ire

Profil

Echelle du Plan

5 10 20 - 40 60. 120. *Toises*

Echelle du Profil.

1 2 3 6 *Toises*

P A

B C C C D P

Plate 78   The Art of War VII

Here we move from defence to offence, still under the guidance of Vauban. For this reason the conception of tactics remains essentially sculptural. Siegecraft consisted of surrounding the defending citadel (A B) with what amounted to another fortress, a light flexible one that could be tightened like a noose. Fig. 1 is a section of the siege line which Vauban threw around the citadel of Philipsbourg in 1676. The small letters refer to the plan of the line and the capitals to the profile shown below (P P). The principles are evidently those of a mobile fortress: the line shelters troops behind a parapet (a A), protected by a ditch (C B), itself secured by a pock-mark band of pitfalls (b C) and a forediteh (d D).

Fig. 2 illustrates the herringbone scheme of attack by parallels. The attacking force begins at the first parallel (a a) beyond range. Approach trenches are dug forward in zigzag fashion, and a second parallel trench (X X) is excavated, which now serves as

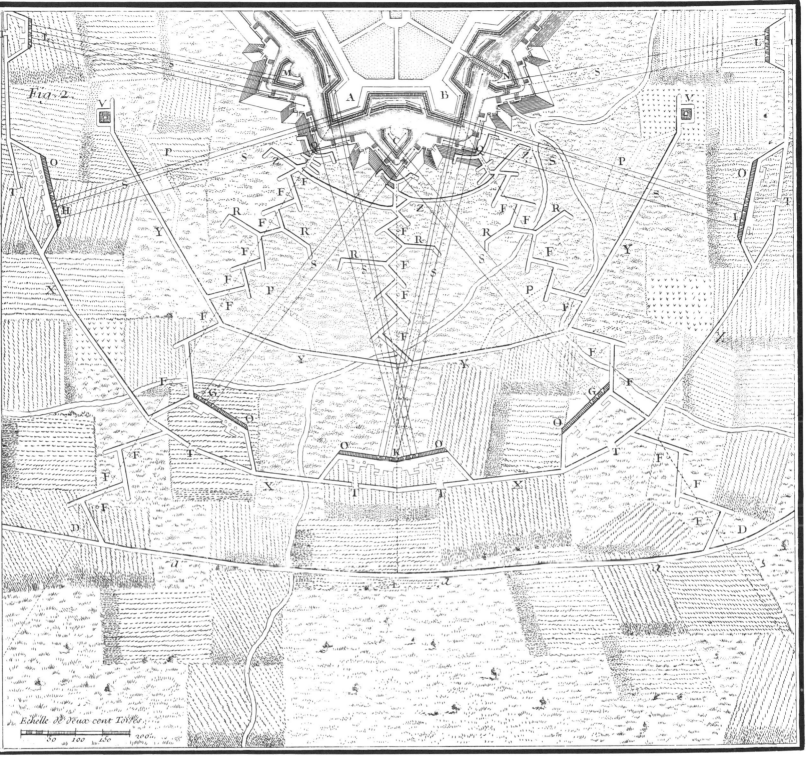

a new base of attack and place d'armes, or weapons supply point. Artillery and mortar batteries (G O, K O) are dug in at suitable emplacements. And so the attack proceeds, like the lumbering charge of some tortoise in his armor, slow but safe, relying not on dash but on patience and attention to the terrain.

Plate 79

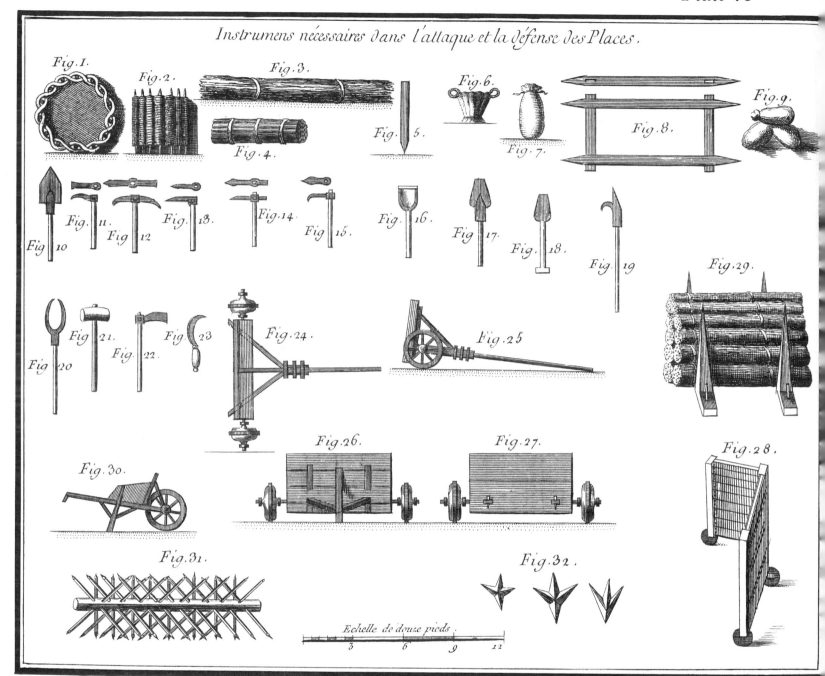

Instrumens nécessaires dans l'attaque et la défense des Places.

Fig. 1. Fig. 2. Fig. 3. Fig. 6. Fig. 9. Fig. 4. Fig. 5. Fig. 7. Fig. 8.

Fig. 10 Fig. 11. Fig. 12 Fig. 13. Fig. 14 Fig. 15. Fig. 16. Fig. 17. Fig. 18. Fig. 19 Fig. 29.

Fig. 20 Fig. 21. Fig. 22. Fig. 23 Fig. 24. Fig. 25 Fig. 28.

Fig. 30. Fig. 26. Fig. 27.

Fig. 31. Fig. 32.

Echelle de douze pieds.
3   6   9   12

Plate 79   The Art of War VIII

Like all good military engineering, 18th century techniques of sapping made maximum use of standardized and prefabricated structures which could be slipped into place with minimum exposure of personnel.

The tools of the trade are shown in the upper section. Figs. 1 and 2 are top and side views of the gabion. Placed side by side, they formed the scaffolding of the protective ramparts behind which assault trenches were burrowed forward. The body of the ramparts was made of bundles of twigs or faggots (3, 4, 29), onto which dirt would be piled. But even so, someone always had to go first, and the French corps of Engi-

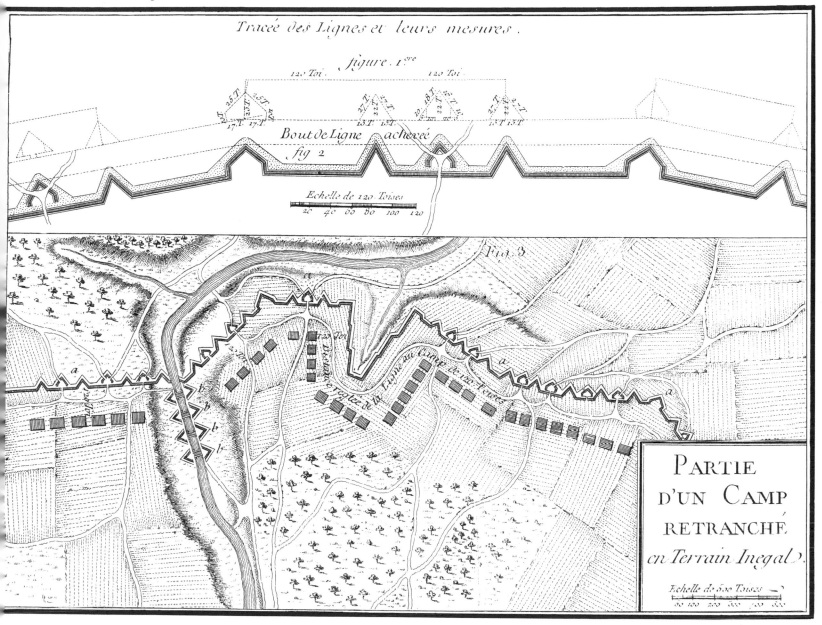

Tracée des Lignes et leurs mesures.

figure 1ère

Bout de Ligne achevée

fig 2

Echelle de 120 Toises

Fig. 3

PARTIE
D'UN CAMP
RETRANCHÉ
en Terrain Inegal

Echelle de 500 Toises

neers had devised a rampart on wheels, a mantelet *(Figs. 24-27)*, which the advance sapper pushed ahead of him. Behind it he could open the trench in relative security. The use of the other instruments is obvious: sandbags *(7, 9)* and a framework for them *(8)* — a blinde, *(from which comes the French word for "tank")*; an early form of mantelet *(28)*; the cheval de frise, *the evolutionary ancestor of barbed wire (31)*; and nasty barbs *(32)* to be scattered about as a protection against cavalry attack.

Above, (Figs. 2 and 3), is a plan for a fortified camp in which a commander might secure his forces if he had to quarter them in the field for considerable periods. It was considered best to select uneven terrain presenting features that could be exploited to provide defensive security, for all aspects of tactics were dominated by the fortress philosophy.

Plate 80

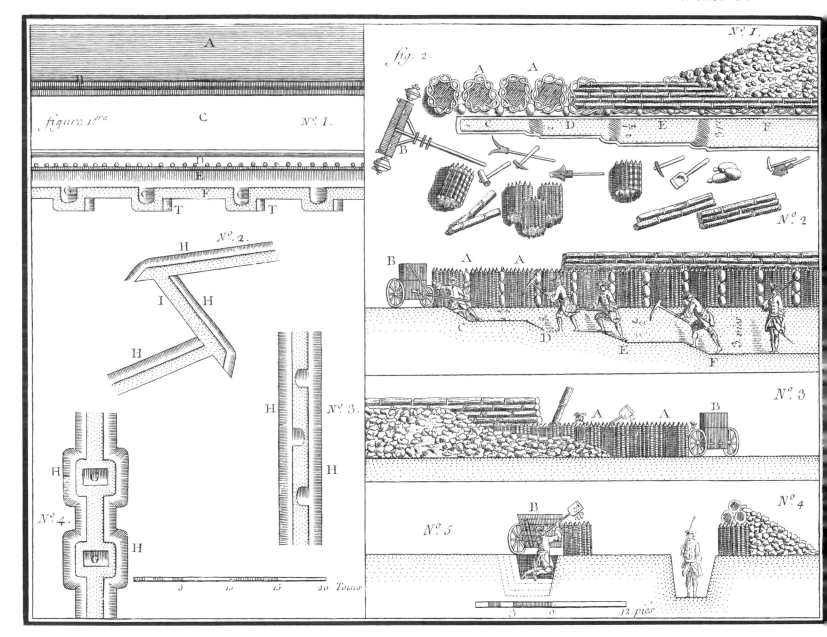

Plate 80   The Art of War IX

The sappers' techniques were as formalized as the plan of attack. This plate illustrates the elements of sapping. Fig. 1, No. 1, is a top-view of the outer defenses of the fortress under attack: ditch (A), contrescarpe (B), covered way (C), parapet and firing posts (D, E, F, T.) Below are approach trenches — the zigzag trenches (No. 2) protected on one side (H), the communication trenches (3, 4) on two.

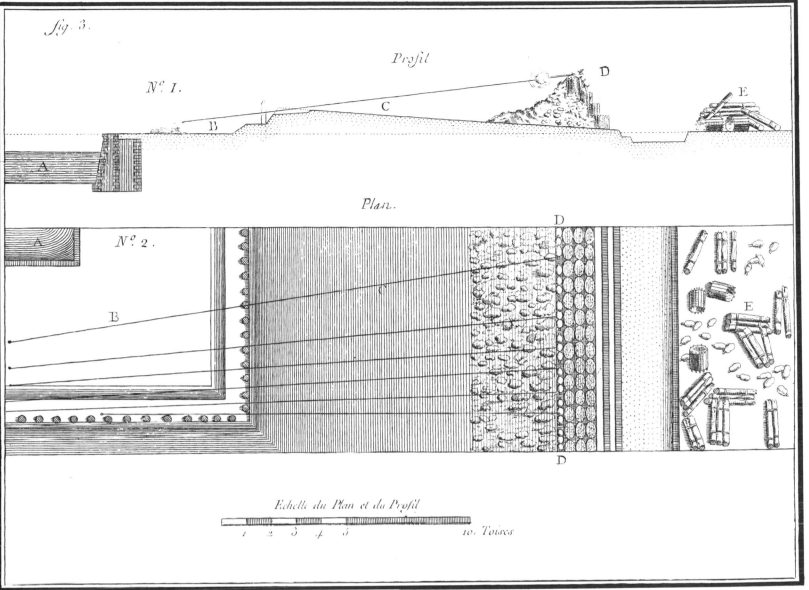

Fig. 3.

Profil

N.º 1.

D

E

C

B

A

Plan.

D

A

N.º 2.

B

C

E

D

Echelle du Plan et du Profil

1 2 3 4 5   10. *Toises*

Fig 2 shows the construction of a trench and its parapet. The ditch is excavated in four definite stages (No. 1 and No. 2), and the parapet is built up against a scaffolding of gabions (A) and logs onto which the dirt is piled (No. 5). Notice that the enemy is given no target more tempting than the wheel-barrow (B).

In Fig. 3 the parapet of an assault trench has been built up several stages farther, high enough for the musketeer (D) to prevent the defender from using the covered way (B). No. 2 shows the plan of this raised parapet — called a cavalier de tranchée *— and of a corner of the outworks it overlooks. (A is the defender's ditch and B his covered way). Sieges were conducted at close quarters; the distance between the two parapets is only 25 yards or so.*

Plate 81

Plate 81   The Art of War X

The siege lines of Fig. 1 have reached the fortress, and now the sappers turn moles and tunnel under the walls to open the way to a breach in the ramparts (B, d). This may be accomplished either by artillery or by the use of mines.

If the defending ditch was empty, the best method of attack (A) was to tunnel under the glacis and contrescarpe and build a barricade across the ditch to the breach (d). The sappers' work was covered all the while by musketry (m). The view (B) is that of the besieging commander looking towards his objective from his own lines. But the ditch may be a moat full of water (C). In this case, the approach is made on the

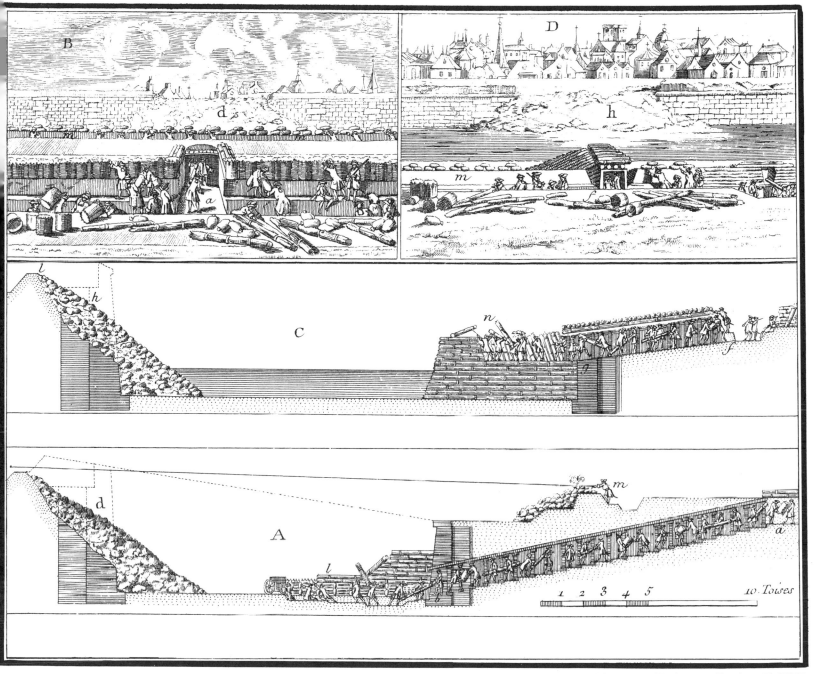

surface, à l'oeil ouvert *(by an open eye), and a foundation thrown down in the moat to enable the attackers to cross. The final view (D) gives the attacker's perspective on this operation. He may not need to go any further. By this time a sane defense would have surrendered.*

Iron Foundry & Forge

Iron Foundry & Forge

It is characteristic of the respective styles of France and England in technology that the great metallurgists of the 18th century should have been French and the great inventors English. Réaumur's *Memoirs on Steel and Iron* have recently been translated from the French and provided with a critical introduction by a distinguished metallurgist of our own time, Cyril Stanley Smith, who writes of their importance in the history of science. But despite the play of reason, in Réaumur and later in the *Encyclopedia,* the iron industry remained relatively backward in France. It was the English ironmasters whose innovations permitted that expansion of the industry which now sustains the whole weight of industrial culture. For it was the 19th century which was truly the iron age. Only then did iron replace wood and stone as the basic structural material.

Iron Foundry & Forge

The history of iron manufacturing is one of steady evolution in capacity and scale rather than of some crucial revolution in technique. The essential invention was permissive rather than causative. In 1709 Abraham Darby, of a famous Quaker dynasty, successfully substituted coked coal for charcoal in the blast furnace. The threat to the woodlands posed by the iron industry—and with it the glass industry, (see Plate 227)—alarmed governments which had to consider both the needs of the Navy and of the population at large. Ineffectual attempts to prohibit the destruction of forests by ironmasters appear on the statute books of both France and England from the 16th century. In England the disappearance of timber had assumed the proportions of a crisis by the 18th century. Coke saved not only the forests, therefore, but the iron trades. Or rather, it did so when its use became known. For Darby was a good Quaker. He kept his own counsel and looked to his own profits, and it was not until near the end of the century that coke was in general use.

In France ironmasters were better supplied with woodlands and not anxious to disturb themselves by adopting new techniques. Indeed, they were often landowners and proprietors of forests, making iron on the side. The first blast furnace to use coke went into operation in 1782 at Le Creusot. It was the nucleus of the great Schneider-Creusot combine, but the portent is apparent only in retrospect, for by 1815 it remained one of only two installations burning coke. This illustrates a curious strain of conservatism running through French industrial history, a conservatism which seems at odds with the progressive theoretical instinct. But, however inconsistent in practice, these attitudes merge psychologically in a peculiarly French form of technological snobbery. Content with its command of principles, French enterprise often resists innovation in the name of quality, and French ironmasters refused to shift to coke and to expose their workmen and their product to its poisonous and adulterant fumes or themselves to the trouble of large scale production. They preferred their little businesses.

Accordingly the *Encyclopedia* contains no hint of the newest methods. And if this diminishes its claim to speak always with the voice of progress, at least we are given a view, somewhat idealised, of the old iron industry at the last possible moment. To get the scale in focus the American reader should think back from the giant plants of Bethlehem or Pittsburgh to the ruins of colonial foundry and forge which are still to be seen in many places in the eastern states — particularly in Pennsylvania. But except that malleable products were rolled or forged out of wrought iron rather than steel, the stages of production were much as now. Ore was smelted in the blast furnace to produce pig iron or direct castings. Pig iron could then be remelted in a foundry to produce cast iron products, which were hard but brittle, or fined in a forge to be wrought into goods made of malleable metal. Steel, which combines the properties of hardness and malleability, was not much used. Its production was so expensive that it served only for cutlery and such luxury items, and it had to await the mid-19th century and the Bessemer and open-hearth processes to displace structural iron.

Plate 82

Plate 82   Iron Mining I

The vignette at the top shows several types of pit-head. The simplest (Fig. 1) is nothing but a manual windlass for hoisting baskets of ore (Fig. 2) from the workings opposite. The construction (above, right) is more elaborate. The hoist is horsepowered, and the yield correspondingly greater. But it was not worth going to the expense required unless the vein was a rich one and the installation permanent. A large proportion of ore was raised by the simpler methods at the left, above.

The right-hand side of the facing vignette takes us down into the mine beneath the pit-head. One miner (Fig. 5) splits a boulder of ore-bearing rock with a wedge, and

Iron Mining I

a second wheels a barrow of ore to the foot of the hoist (ef), where it will be dumped into baskets (d). But this illustration is misleading. It makes a mine look cleaner and lighter than any 18th century mine ever was. Actually, the shafts were dark and noisome, damp and dangerous. Often springs of water ran right through the galleries. Lateral shafts would be driven out the hillside to carry off the water—as at (o). Nor were they all shored up as carefully as is the right-hand gallery (g).

Given his choice any miner would have preferred employment at the surface in a strip-mine like the one at the left. Not only were the conditions more agreeable, but gunpowder could be used to save labor without the risk of burying the miners in the debris of the mine. The miner sitting down to his work is drilling a hole with a diamond-pointed drill. A charge of gunpowder will be poured in and the rock blasted loose to be broken up for ore.

Plate 83

Plate 83   Iron Mining II

The top illustration shows iron ore at its most accessible, in surface deposits of a gravelly consistency. The miner loading the cart (Fig. 3) has only to reject the non-ferrous rocks. In certain parts of France, particularly in Brittany, the swamps and shallow lakes were rich in iron pyrites, which were, for the small-scale operations of the time, well worth reclaiming (at the right).

Iron Mining II

Plate 84   Iron Mining III

Plate 84   Iron Mining III

*Most ores, "earthy" ores, had to be washed before being smelted. This simple oper-
ation is here carried out in a series of basins located below the dike of a canal. The
basins draw their water through two little sluices (d, c) and underground conduits
of which the course is indicated by dotted lines. One method involves washing the
ores in a basket (Fig. 1), and some workmen save their backs by rigging up a sort of
flexible crane (o) on which they suspend the basket. The more usual technique is
simply to shovel the ore into the basin (Fig. 2), stir it about (Fig. 3), and drain off
the wash-water (Fig. 4).*

Plate 85   Iron Mining IV

Plate 85   Iron Mining IV

In large installations the washing of ore might be mechanized. Various devices were in use. At the left, the ore is shovelled into a hopper which drops it into a swift stream of water flowing over an inclined lattice (A). The ore falls through the grille with the water, and the pieces of rock are screened out to tumble off onto the ground. A workman (Fig. 3) stirs the ore for further washing in the basin. The large machine at the right would have been used only in fairly elaborate and well-capitalized establishments. It is a water-powered agitator. A suspension of ore and water is stirred in the trough (HH) by the rotation of the rectangular iron bars (R) affixed to a shaft (N). The mixture is discharged into a settling basin (S).

Plate 86 The Blast Furnace I

Here is a general view of a blast furnace and its surrounding buildings. In choosing a location an ironmaster had first to consider the availability of water power and fuel. He consumed such prodigious quantities of charcoal that its transport was far more expensive than that of ore. It was, indeed, prohibitive.

The stream in the illustration runs strongly to turn his wheel, while on the opposite bank horses bear charcoal to his warehouse (PP) from some nearby woodland. On the upper platform the charger (Fig. 1) is gauging the furnace. If the top of the charge is three feet below the throat, it is time for another feeding. Notice the open window (A) and skylight through which the men at the hearth and the throat of the furnace can shout back and forth.

Outside the door a workman (Fig. 5) weighs pig No. 289. A furnace like this might produce two tons of iron a day. From the ironmaster's point of view there is one jarring note in this scene: the man (Fig. 4) in a cocked hat with a goosequill. He is the clerk representing the lord of the manor, and he writes down the weight and number of each pig in a register from which are computed the dues owing to the domain. In England no such functionary would exist. Freedom from feudal obligations was one of the major advantages enjoyed by the fortunate manufacturers of the English Midlands.

Plate 86   The Blast Furnace I

Fig. 1.

Fig. 2.

Fig. 3.

Fig. 4.

Fig. 5.

Plate 87 The Blast Furnace II

"A blast furnace," writes the Encyclopedia's expert on iron, "is really a stomach which demands feeding steadily, regularly, and endlessly. It is subject to changes in behavior through lack of nourishment, to indigestion and embarrassing eruptions through too rich or voluminous a diet, and in such cases prompt remedies are to be applied."

The plate gives two right-angle cross-sections of an 18th century blast furnace. It differs from a modern blast furnace in scale and detail, but not in principle. The height of the shaft (EL) was here about 25 feet. It consumed charcoal instead of coke. The blast was supplied by a water-powered pair of bellows (side-view, Fig. 2, RR) blowing alternately. Notice the counter-weights (i, ii) on the bellows shafts.

Charcoal, iron ore (washed, dried, and pulverized) and a limestone flux are fed into the throat (L). The furnace is kept full and so designed that the charge will always be descending, sustained by the outward flare of the boshes (EI), and settling very gradually as charcoal is consumed in flame and as molten iron and floating slag are drawn off from the hearth (E). Combustion is maintained by the forced blast of air rushing up through the descending mass—the article in the Encyclopedia sustains the metaphor and speaks of the "digestion" of the materials in the "body" of the furnace.

In principle and origin a blast furnace is only an elongated forge, a forge turned giraffe. But simple as it was, its operation was subject to all sorts of accidents. The tuyères—the vents through which the blast is blown—might become clogged. Clinkers of vitrified slag might choke the hearth, and if they grew big enough—as could happen very suddenly—the furnace might have to be stopped and dismantled. To prevent this, the hearth was covered with the finest sand. The furnace might burn out and collapse before its time. The ordinary life of a furnace was 30 weeks, after which it would have to be stopped, emptied, and relined. The more often this happened, the higher was the overhead. An alert foreman would foresee miscarriages, and if he did so in time, would shut up the furnace temporarily to forestall loss and disaster.

Against the danger of explosion, however, there was no defense. "For workmen and plant alike, eruptions are the most terrible danger. They bring death to those nearby and spread fire far and wide. In a sudden explosion, a furnace will throw up all its contents, molten and solid. It becomes a volcano vomiting flaming fragments from every opening." Such disasters were caused by stoppages in the venting, or by a charge which had settled unevenly and formed pockets in which explosive vapor accumulated. The symptom was a flame which, instead of burning evenly, alternately disappeared and leaped abnormally high. If the workers should see the furnace gasping in this way, "flight is the most expedient measure."

Plate 87

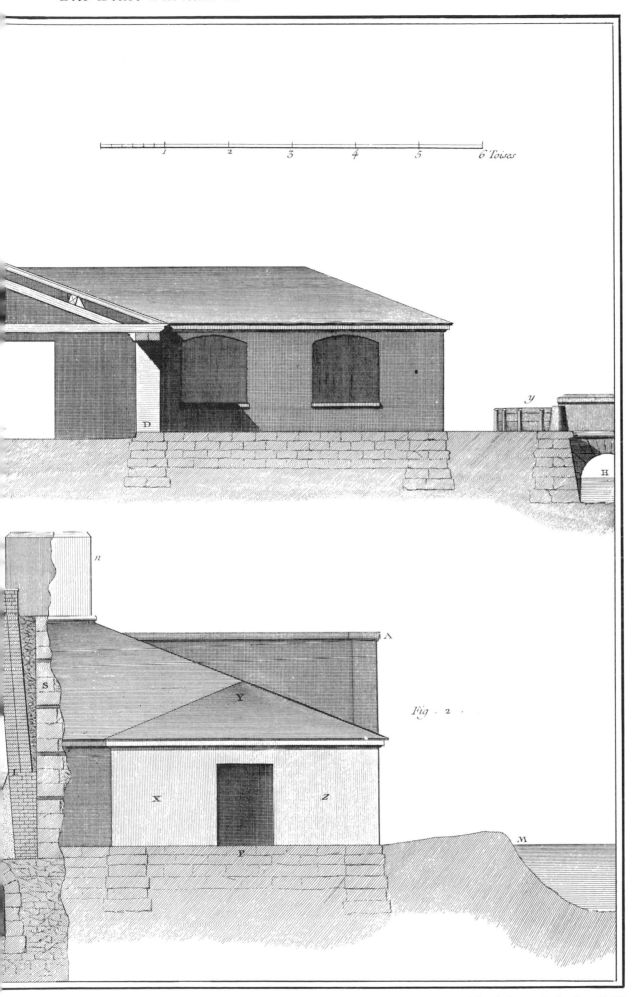

Plate 88 The Blast Furnace III

In certain foundries in Dauphiny the draft, instead of being supplied by a bellows, was created by the ingenious pump illustrated in this plate. Its advantages are that it has no moving parts and that the draft is constant. The principle is the same as that of the suction pump attached to water faucets in many modern chemistry laboratories, and student chemists who have been incautious, only to see their filter paper ruptured and their precipitate disappear up the hose, will appreciate the force of which the arrangement is capable. The internal construction of the fallpipes is shown in Fig. 2. A constriction (E) in the pipe just below the reservoir inlet (C) narrows the flow and increases its velocity, with the result that air is constantly sucked in through the "sigh-holes" (D, d) located at the point (e) where the pipe resumes full diameter. The air is carried down in a foamy fall into the barrel below, where the water, kept at constant level (K) by a sluice (N), splashes off an iron plate (H), while the air is forced out (p) into the blast pipe (mn). Three barrels are vented together to add volume to the draft, which is given velocity by the nozzle (u) applied to the tuyère of the furnace.

Plate 88   The Blast Furnace III

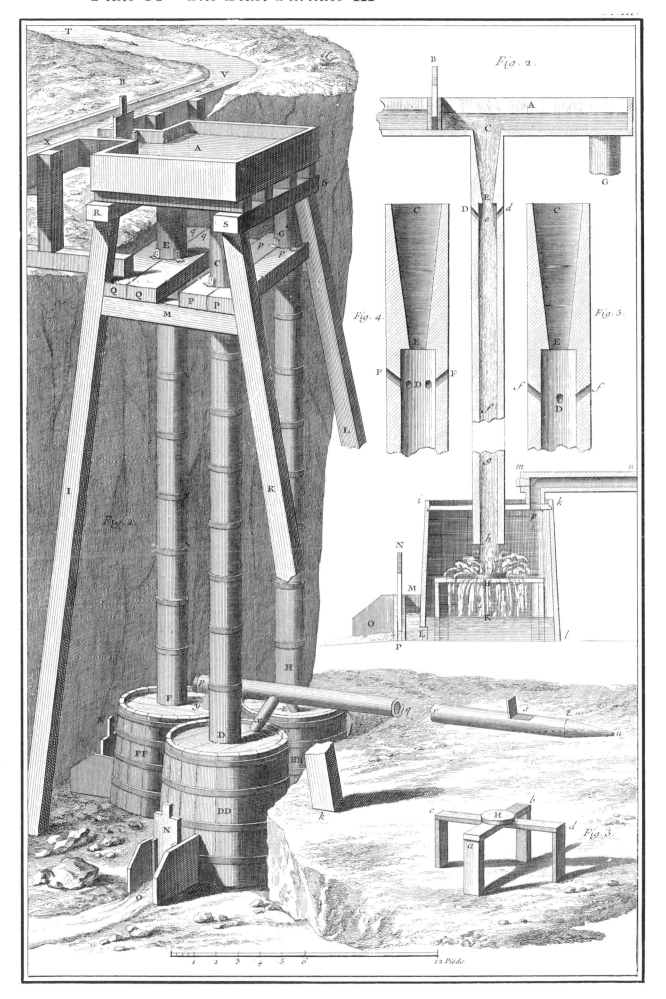

Plate 89   The Blast Furnace IV

The furnace was replenished when the mass had settled enough to make room for an added charge. This consisted of 230 pounds of charcoal, 500 of ore, 50 of limestone, and 20 of argillaceous earth as a sort of protective lubricant. These ingredients had to be added in a certain order. Three baskets of charcoal, half a basket of limestone, and two more of charcoal, were placed to give a surface tilted at an angle of 30°. On this were charged ten baskets of mineral. The tilt prevented the mineral from settling straight down through the center, or from crushing the charcoal. These practices were strictly enforced, and in this furnace the chargers were required to keep ten stones with which to tally the baskets of minerals. The total capacity of the furnace was about 7200 pounds. A single charge would move through the blast in 12 to 14 hours, and in a good week the furnace would produce 6 or 7 tons of pig iron.

With good luck and management a well-built furnace could be kept in continuous blast for 20 to 30 weeks, after which it would be "blown out"—i.e. emptied for repairs to hearth, lining, and bellows. New reserves would be accumulated of iron and charcoal, particularly the latter, which always presented a critical supply problem. Resuming operations, or "blowing in" the furnace was no simple matter. It had first to be filled with charcoal, which was ignited without the blast. The fire was allowed to work its way up to the throat in order to heat the furnace gradually. Three days would elapse—the ironmasters were generally inclined to be impatient and to shorten this interval more than the Encyclopedist thought wise—before the furnace was ready to receive iron ore. It would be two or three weeks before the correct proportions were attained throughout the entire charge, and during this period only gray iron of poor quality could be expected. It was no light matter, therefore, to interrupt a blast to correct errors or prevent accidents.

This view of the top of the furnace shows the throat (E) and above it the hood and chimney which carry off the flame and smoke. The sides are cut away to afford a view. The beams (e, e) are the back ends of the bellows shaft in their housings (k), each with a counterweight (i). It was necessary to harness a pair of bellows so that one might blow while the other filled. A steady blast was of great importance, and maintenance of the bellows was as responsible a task as charging the furnace.

Plate 89   The Blast Furnace IV

Plate 90   The Blast Furnace V

The iron smelted in the blast furnace was destined for either of two main divisions of the industry. An establishment which was at once furnace and foundry would produce cast iron products like pots, chimney-backs, and cannon directly on tapping the furnace. But the larger proportion of the iron smelted was cast into "pigs"—great rough bars of crude iron—to be sold unfinished to foundries and forges for fabrication into the products made of malleable iron.

To cast a pig, molten iron was tapped from the hearth (I) into a mold prepared in a bed of sand. The furnace master and his assistant make the mold, or rather remake it, while two other workmen maneuver a pig which has just hardened and cooled. In order to keep a production tally, each pig is numbered by a system of modified Roman numerals molded right into it. The pig on the rollers is number 287 of the blast. Notice the bellows at the left (R, RR). When the furnace is being tapped and the hearth gate is open, the blast must be arrested. Otherwise the workmen "who are already only too exposed to great heat," would be suffocated by clouds of flame and fumes. It is a defect of the quality of these plates, as of the mining plate above, that their clarity makes a blast furnace look more like a laboratory than like what it was: a grimy, sweaty little inferno.

Plate 90 The Blast Furnace V

Fig. 1.

Fig. 2.

Fig. 3.

Fig. 4.

Plate 91   The Blast Furnace VI

When the iron runs in the hearth and the mold is ready, the furnace master (Fig. 1) pulls aside the hearth-gate. Tapping was described by an English commentator of the late 17th century. (English practice differed from French mainly in casting the iron into several molds branching off the main one called the Sow*):*

"They can run a Sow *and* piggs *once in 12 hours, which they do in a bed of sand before the mouth of the* furnace, *wherein they make one larger furrow than the rest next the* Timp *where the metal comes forth) which is for the* Sow, *from whence they draw two or three and twenty others (like the* labells of a file in Heraldry*) for the* piggs, *all which too they make greater or lesser according to the quantity of their Metall: into these when their* Receivers *are full they let it forth, which is made so very fluid by the violence of the fire, that it not only runs to the utmost distance of the* furrows *but stands boiling in them for a considerable time."*[1]

The molten iron issued from the furnace with slag floating on the surface. Skimming it off was the function of the bar G laid across the mold. Some iron masters preferred to protect the metal from the air as it blackened and solidified. As soon as the tap was completed, a workman (Fig. 2) tossed a layer of ashes across the still red surface. French technological imagery differed from English. Instead of a sow and pigs, this furnace produced a gueuse — slut. *The device (F) for closing the hearthgate was a* dame. *It was designed so that the bulk of the slag would run out over it while the metal was retained in the receiver.*

1 Natural History of Staffordshire *by R. Plot, quoted by T. S. Ashton in* Iron and Steel in the Industrial Revolution. *(Manchester, 1924), p. 233.*

Plate 91   The Blast Furnace VI

Plate 92   Cast Iron I

Eighteenth century iron falls into the two main categories of cast iron, which was hard, brittle, and unworkable, and wrought iron, which had to be refined at the forge to make it malleable. Cast iron might be made in a separate foundry, but it was technically possible to cast certain objects right from the blast furnace, and if the furnace was near a good market, it might pay to divert some portion of its output from the production of pig. Furnaces arranged to cast directly on tapping were known in France as Fourneaux en marchandise *(merchant furnaces), as opposed to the* Haut Fourneau *which ran pigs. In English this distinction has been transferred to the fabricating departments of steel plants, where the "merchant mill" is differentiated from the rolling mill producing heavy rails and girders.*

There were many techniques for casting iron. Casting in molds of sand gave the finest surface. This was easy enough for flatwork like chimney-backs and ornamental iron. It presented greater difficulties in casting hollow objects—skillets, kettles, water conduits—for the mold was apt to crumble while being filled.

The principle of casting in sand was simple enough. It consisted in filling a wooden form (Fig. 1) with damp sand (m)—of about the consistency of the sand which children use to make castles on a beach. The sand was molded around a model of the utensil to be cast—three-legged cauldron, iron pipe, or whatever (Fig. 2). The model is removed, the form closed, and molten iron cast into the mold (Fig. 3).

Flat objects, like the chimney-back leaning against the rear wall, might be cast right in the sand of the foundry floor. A workman (Fig. 4) is pressing a wooden form bearing the design into the floor, and the iron will be run into it in the way that English foundrymen ran a litter of pigs off the sow. (The taphole of the furnace is just to the left, in the bay between the pillars ST).

Cast iron products inevitably exhibited imperfections. There was bound to be a rough spot at the point where the iron was poured when molten, and this had to be smoothed over with file or hammer on a sort of two-legged sawhorse (Fig. 6).

Some objects were better cast in earthenware molds than directly in sand. The molds were generally not strong enough to support their contents of molten iron. They might either be buried in the floor or packed in sand in the same forms used for sand casting.

Plate 92   Cast Iron I

Plate 93   Cast Iron II

Plate 93   Cast Iron II

This workman (Fig. 1) is making an earthenware mold on a potter's wheel. His clay (B) is beside him, and in the back of the shop are ovens (q) where his molds are baked and cooled (ef). The one on the right (p) is cut away to show the simple construction.

Plate 94   Cast Iron III

In this establishment earthenware molds were baked right in the foundry, to the right of the furnace (q, p). Whether the molds were of clay or of sand, they were filled in one of two ways. If embedded in the floor, they could be filled by tapping the furnace

Plate 94   Cast Iron III

(Y). But to tap a furnace for small quantities was risky, and it was generally preferable to ladle molten iron from the hearth (Fig. 1) and to cast by hand (Figs. 4 and 7). The two boys are skimming off slag. But there must be some convention about their presence, for it seems to be a feature of 18th century technical iconography that the ladling of molten materials involves a gamin to attend the ladle. (See, for example, Plates 224, 226 on the glass industry). The section of pipe in Fig. 8 is a typical example of the sort of product to which cast iron was suited. The workman is chipping off the scale that remains from the mold.

Plate 95   Wrought Iron I

Production of wrought iron was the second great division of the industry in the 18th century, for the use of which crude pig iron had to be fined and rendered malleable at the forge. The most common example of the forge was the village blacksmith's shop, and the history of the iron industry shows no industrial revolution, but rather a steady expansion in quantity and scale of production. The forge shown here, as impressive an installation as could be found in France, represents approximately a midpoint in the evolutionary progression from the medieval smithy to the steel plant. In another sense, however, it is almost the culmination of one line of development: its product is still iron, whereas the future for malleable metal in the 19th century lay with steel.

The forge itself is an open hearth, fired with charcoal like the blast furnace and fanned by a pair of water-driven bellows rather smaller than those for the blast. It is a very hot place. The men work face to face with an open fire, manipulating and beating great lumps of heat-soaked iron, and they must wear protective clothing: leggings, heavy workshirts and aprons, broad brimmed hats to shield face and eyes from the fire. Pig iron is loaded onto the hearth through the opening at the back (7), and a forger (Fig. 1) shoves the pig bit by bit towards the flame as the softened portion is collected (Fig. 2) into a "loop." Impurities, the "scories", run out from the bottom of the hearth.

Fining iron consists essentially of beating out impurities until it assumes the desired malleability. The process begins in Fig. 3 where the "loop," having been removed all pasty from the forge, is hammered on an iron plate into a roughly rectangular "half-bloom," suited to the blows of the tilt-hammer at the right. The operation of the tilt-hammer will be clearer in the following plate.

Plate 95   Wrought Iron I

Plate 96

Plate 96   Wrought Iron II

Conversion of the pasty "loop" into bar iron fit for manufacturing required four or five heats and forgings. The successive forms into which the iron is worked by the tilt-hammer are shown at right: the "loop" (Fig. 5), the "half-bloom" (Fig. 6), and the "bloom" (Fig. 7), with its two ends, the "ancony" and the "mocket head." In Fig. 8 the "ancony" has been beaten out to a bar, and in Fig. 9 the "mocket head" has been flattened.

Between each of these stages the iron had to be returned to the forge for reheating, and large establishments usually kept two forges busy at the work, one, the "Finery," where the bloom was formed, and the other, the "Chafery," where it was forged.

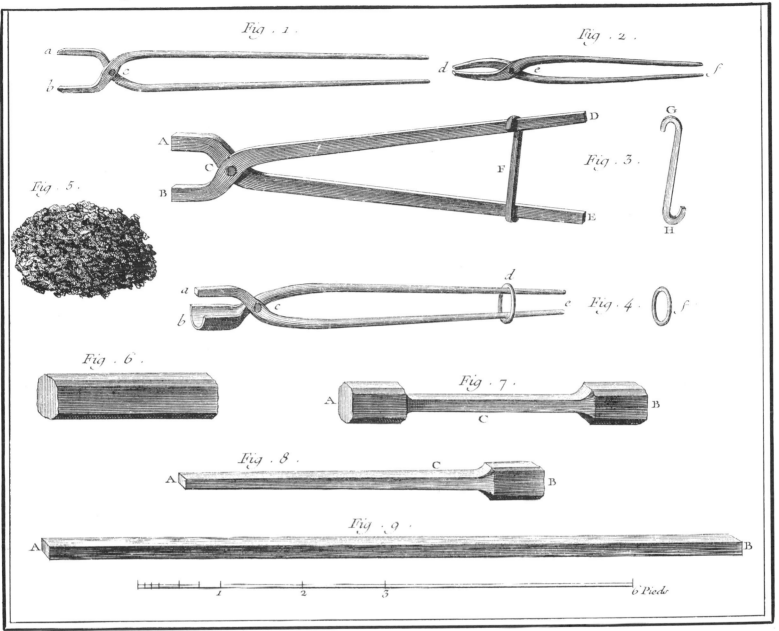

Fig . 1 .

Fig . 2 .

Fig . 3 .

Fig . 5 .

Fig . 4 .

Fig . 6 .

Fig . 7 .

Fig . 8 .

Fig . 9 .

1   2   3   6 *Pieds*

In this plate the forging begins. The forger (Fig. 3) turns the half-bloom (a) from side to side on the anvil (W) between strokes of the hammer (H). After each blow the hammer is raised by a water wheel, which is not shown except for the counter-beam (13) that holds the hammer suspended. A worker behind the forger (Fig. 2) controls the clutch which engages the mechanism.

Plate 97   Wrought Iron III

Plate 97   Wrought Iron III

The cam mechanism for lifting the hammer is illustrated in the forging of the bloom. The great shaft (Y), bound in iron bands, is the axle-tree of a water wheel outside the smithy. When it is engaged, one of the four arms catches the shaft of the hammer under the band (r) and bumps it upward into position for another stroke. As can be imagined, a forge was an inferno, not just of heat, but of noise, what with the creaking and blowing of the bellows, the groaning of this axle (Y), the crash of the falling hammer, the clanking of its rising, and (perhaps) the cursing of the men when, as often happened, things did not go as smoothly as in these pictures.

Plate 98   Wrought Iron IV

Plate 98   Wrought Iron IV

Cold water cast on the bloom (Fig. 2) at the moment of hammering assists in scaling off impurities, and it compounds the banging of the hammer with the hissing, almost the shriek, of water exploding into steam. As the Encyclopedist says, this "makes a noise than which nothing could be less agreeable to the ear." This view shows the hammer being raised. Notice that the shaft has to be protected from the heat of bloom and anvil by an iron shield (behind HK).

Plate 99   Wrought Iron V

The bar which is the goal of the fining process has been so altered in its physical properties that it can be wrought into whatever forms a manufacturer may desire.

Here is a combination rolling and slitting mill, which flattens bar iron and slits the strips into rods. First the bar must be soaked to a white heat for at least an hour in the wood-burning oven (Y). Then it is progressively thinned and lengthened between rollers (C, D, V¹u¹ etc.). The gap between the rollers narrows from one pair to the next. In this picture the flattened bar is passing toward the reader through the slitting mill. It emerges as a bundle of rods like those stacked against the wall at the left. A glimpse of the water wheel which powers this operation may be had at the far left. The shaft (E) which runs the rolling mill is turned by a wheel off to the right.

Plate 99   Wrought Iron V

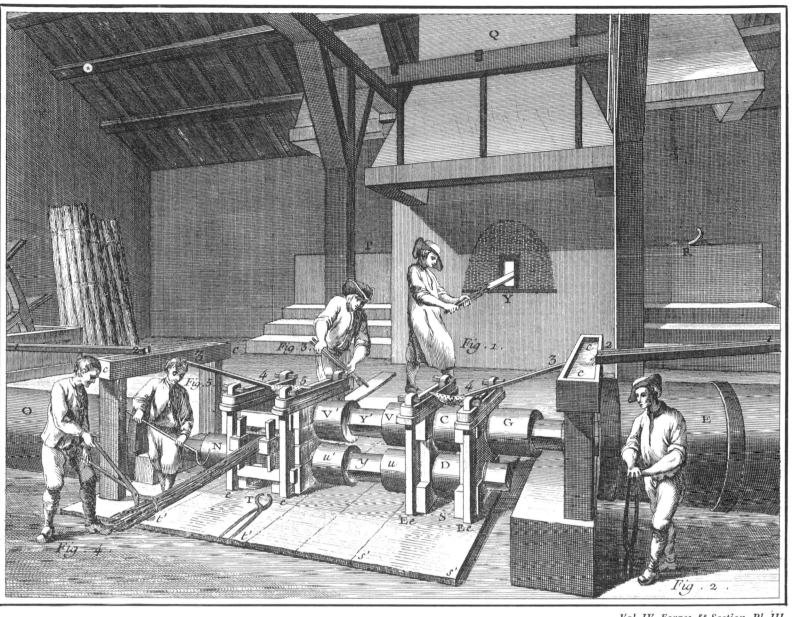

Plate 100   Wrought Iron VI

Fig. 3 gives a schematic top-view of the construction of the roller (SS) and the slitter with its teeth (TT). The machine at the top (Fig. 1 with detail in Fig. 2) rolls a flange into the bar moving through it from A to B.

Plate 100   Wrought Iron VI

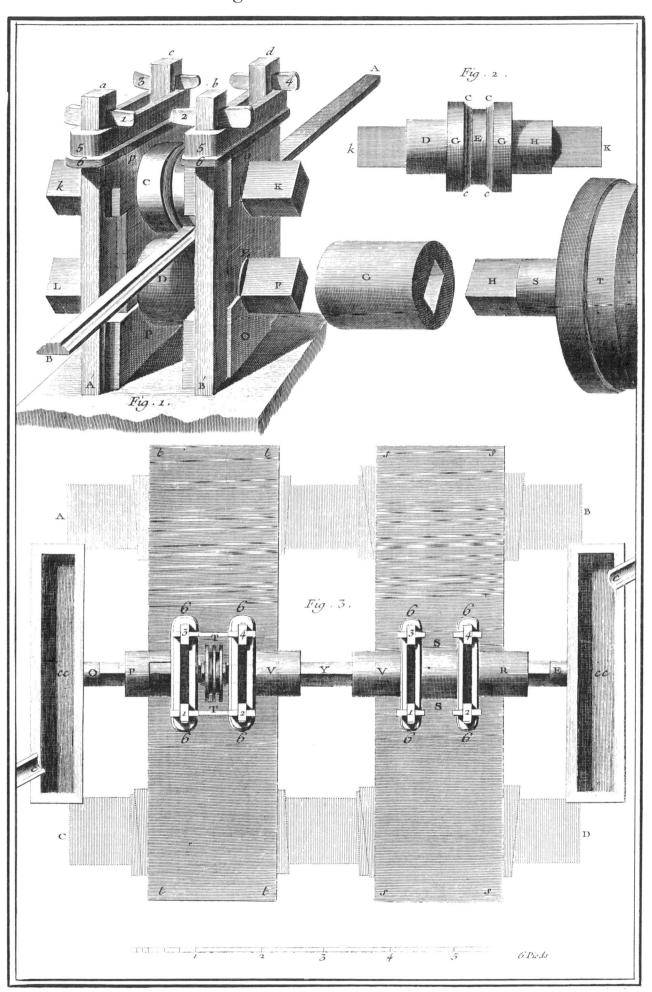

Fig. 2.

Fig. 1.

Fig. 3.

6 Pieds

Plate 101   Wrought Iron VII

Plate 101   Wrought Iron VII

Two workmen bind rods into bundles for shipping. They use the little forge in the background for working the bands.

Plate 102   Forging an Anchor I

Plate 102   Forging an Anchor I

Besides bar iron destined for milling, the 18th century forge also produced items too bulky to be handled by the available machinery. In such cases the forge was essentially an overgrown blacksmithy. Marine anchors, for example, were forged by hammer and manpower. This plate shows the shaft of an anchor, destined for some capital ship, being forged under the triphammer. It would take up to eight or ten heats to bring the bloom to this stage.

Plate 103   Forging an Anchor II

Plate 103   Forging an Anchor II

A claw is being welded to the anchor arm. The masterforger (Fig. 1) supervises very closely. He uses his rule to show where he wants the blows to fall. These are delivered in turn by the four forgers with hammers, while a fifth (Fig. 2) holds the iron so that it will not jump on the anvil. Joining pieces in this fashion was the hottest work of all, for the metal had to be kept as close to red heat as possible while blows were rained upon it.

Plate 104   Forging an Anchor III

Plate 104   Forging an Anchor III

*Before the arm can be joined to the shaft both must undergo strong heat, and must
be brought together as hot and as fast as possible. This means two hearths, the one
at the right for the arm and the one in the background for the shaft. Both these great
parts are suspended from cranes pivoted to swing them together on the anvil. Even
as they turn, the forger (Fig. 7) pulls the clutch which turns water to the wheel, and
knocks out the support (Q) on which the hammer has been resting. "The most violent
blows imaginable are struck as rapidly as possible . . . so that the arm is welded to the
shaft in less time than it takes to read of it."*

Plate 105   Forging an Anchor IV

Plate 105   Forging an Anchor IV

The second arm is joined. Once again the ends to be welded must undergo simultane-ous heats, and the one-armed shaft (OBP) has to be flipped over before it is swung back onto the anvil. The Encyclopedia *does not give weights but it is obvious that the cranes and pulleys are arranged to handle loads of many tons at forging heat. Nor does the masterforger (Fig. 1) allow his men to forget that it will be too late if any weakness in the weld should show up when the claws are dragging in the mud or rocks of the bottom of the sea.*

Plate 106   Forging an Anchor V

Plate 106   Forging an Anchor V

Finally a chisel man (Fig. 1) knocks off the burrs and finishes the rough spots left in the forging. The anvil is placed over a pit which permits the anchor to be worked on from all four sides successively.

Plate 107   Forging an Anchor VI

The last plate shows a different forge, part of a shipyard at some great port where ships are refitted. In such a location there is not likely to be a fall of water sufficient to power a triphammer, and the point of this illustration is to show a hammer worked by a sort of tug of war. Six workmen, three behind the hammer, take the two ends of a rope which passes over the flywheel of Fig. 11 below and by pulling alternately to and fro they run the hammer by manpower.

Plate 107   Forging an Anchor VI

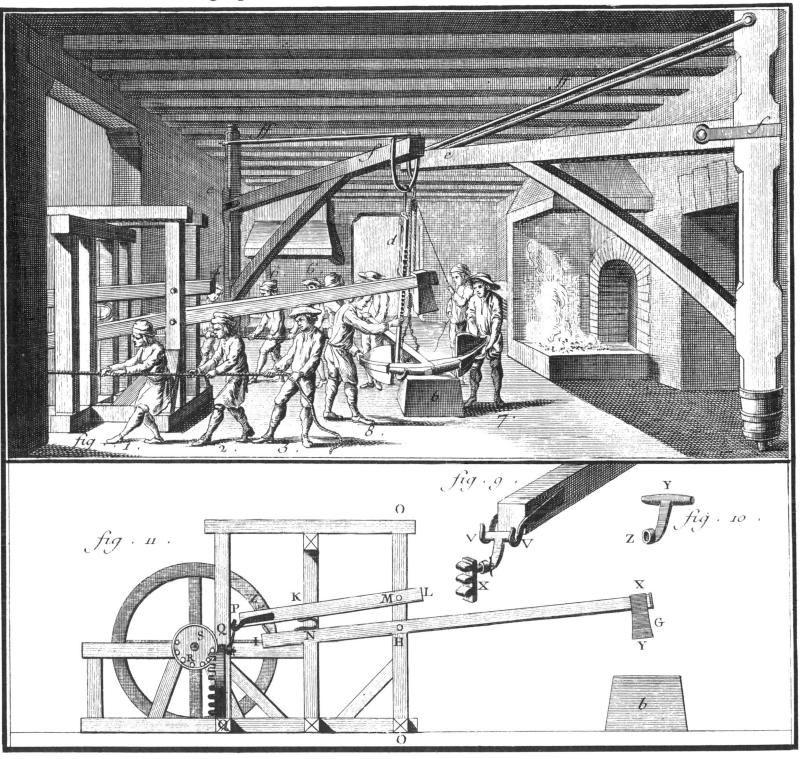

Plate 108 Casting Cannon I

Cannon foundries provided an outlet for a very considerable proportion of the pig iron produced in blast furnaces. Bronze cannon, it is true, were preferred by the army. They were heavier, but if they burst, they would split open in a harmless tear, whereas iron cannon would fly into shrapnel, thus destroying the gun squad as well as themselves. This had the disadvantage, according to Réaumur, the foremost of the 18th century metallurgists, that other gun crews after witnessing such an accident would use light charges, and the balls would fall short of their targets. Nevertheless only the artillery could afford bronze, and naval ordnance continued to use iron cannon, despite their tendency to blow up. Besides being cheaper, they were lighter in weight.

Here are the standard 18th century artillery pieces. From bottom to top they are the 4, 8, 12, 16 and 24-pounders, which designation refers to the weight of the ball.

Plate 108

Plate 109   Casting Cannon II

Plate 109   Casting Cannon II

Casting cannon required three steps: making a clay model of the piece, constructing a mold around the model, and pouring molten iron in the mold. These workmen are winding (Fig. 2) thick rope onto the core (trousseau), which will serve as a foundation for modeling the clay.

Plate 110   Casting Cannon III

Plate 110   Casting Cannon III

A model in clay of the cannon to be cast is built up on the rope winding of the core.

Plate 111

Plate 111   Casting Cannon IV

Trunnions are affixed (Fig. 1) and the embellishment of the cannon imprinted by wooden blocks at the right.

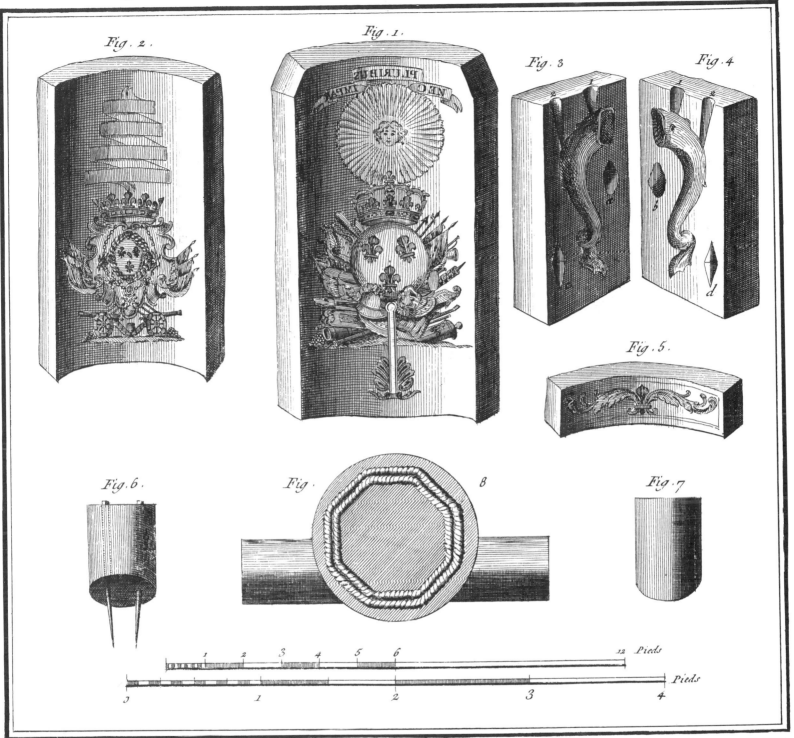

Plate 112   Casting Cannon V

Fig. 1.   Fig. 2

Plate 112   Casting Cannon V

Next the model is covered with a greasy mixture of oil and tallow so that it can be slipped out of the mold which will be formed around it. The mold itself is built up of a different clay and is girdled with iron bands so that in the process of casting it will hold its shape.

Plate 113   Casting Cannon VI

Vol. V, Fonte des Canons, Pl. XV.

Plate 113   Casting Cannon VI

A fire is built right on the workshop floor under the mold to dry the clay. Though shown only once, this operation has to be repeated a number of times, for both model and mold are built up of layers of clay, each of which has to be dried before the next can be applied.

Plate 114 Casting Cannon VII

Casting required skill and judgment. Pig iron or — more usually — bronze gun metal is remelted in the foundry furnace. Impurities are skimmed off by the puddler (Fig. 3), directed by another workman whose attention (Fig. 4) is fixed upon the melt. Cannon molds, two of them in this case, are fixed in place in a pit below the foundry floor. When the furnace is tapped by the masterfounder (Fig. 1), the molten iron surges out into channels which communicate at the end of each lateral transverse with an opening in the mold. Molds are filled one at a time, and the workman (Fig. 5) stops the opening of the one nearer the furnace until the casting of the further one is completed. In theory this foundry serves but a single customer, an august one: the King. Hence the lilies of France over the taphole, and the officers of the Royal Artillery (Fig. 6) who supposedly ensure that specifications are strictly observed.

After casting and annealing, the barrels still have to be trued by boring, an operation usually carried out in the foundry. For an illustration of the boring machine, and of other types of ordnance, see Plate 59, and Plate 69ff.

Plate 114   Casting Cannon VII

Plate 115   Casting Bells I

Variations of the same techniques were applicable to the casting of church bells. The alloy, 75% copper and 25% tin, is fused in a bee-hive furnace at the upper right. The floor is cut away along the lines P and G to let us see down into the casting pit, where (Fig. 3) the outer mold is suspended on a pulley. It can be lowered to fit over the inside form like a hat. The bell will be poured in the space left between the two molds. The workmen (Figs. 1 and 2) ready a mold to receive its charge of molten bell metal.

Plate 115 Casting Bells I

Fig. 3

G

Fig. 2

Fig. 1

Plate 116   Casting Bells II

A single tapping will cast four bells. Two are cast from the transverse trough (UT, ut). Then, when the molds beneath are full, the gate closing off the second trough is raised and the final two bells, which are to strike mi and sol, are cast. At the right a worker (Fig. 4) pushes the last of the metal towards the tap-hole.

The purity and temper of bell metal are all-important. In order to prevent it from solidifying prematurely in the troughs, they are heated by charcoal fires before the tapping, and the wooden tools are set afire before immersion in the melt.

"When it is time for casting, all the canals and passages which have been heated by charcoal fires during the melt are cleaned. All vents are opened. The end of the founder's staff is charred in the fire, and with it the rest of the tools, in order to prevent any spitting of the metal. Everything being ready, the founder (Fig. 1), in slippers and doublet, opens the hearthgate with a great blow of his staff. The metal pours out like a torrent of fire, and without taking his staff from the taphole, he directs the flow to fill the canals. The instant it starts, a flame like that of brandy leaps from the flue and dies down only when the molds are full and the bell casting a success."

Plate 116   Casting Bells II

Plate 117 Casting A Statue I

Bronze was cast even more dramatically in the manufacture of great statues. The most difficult were the martial figures on horseback erected to celebrate the victories of French arms. The first step is to construct a skeleton of wrought iron bands. Bronze for statues would usually contain a higher proportion of copper than would bell metal—up to ninety percent in some cases. In really large statues like this an addition of one percent of brass would make it cast better.

Plate 118   Casting A Statue II

Next the artist creates his statue and builds a plaster cast around the model.

Plate 119   Casting A Statue III

Then he makes a wax impression in the cast. This is the famous statue of Louis XIV erected in 1699 by the city of Paris in the Place Louis-le-Grand. What looks like an external circulatory system is precisely that: a network of channels through which wax will be melted out and replaced by bronze in the casting.

Plate 120   Casting A Statue IV

At the left the mold is cut away to show how it will be filled with bronze. At the right is given an external view of the mold as it awaits the casting, reinforced with bands of iron.

Plate 121 Casting A Statue V

The spirit of the foundrymen is equal to their subject. The taphole will be opened by shoving the barrier into the furnace (5, 6). Molten bronze will pour into the basin (9). When the holes (7) are unstoppered, the bronze will flow down the network of channels prepared to fill up with eternal bronze the hollow form of Louis the Great and his horse.

Plate 117

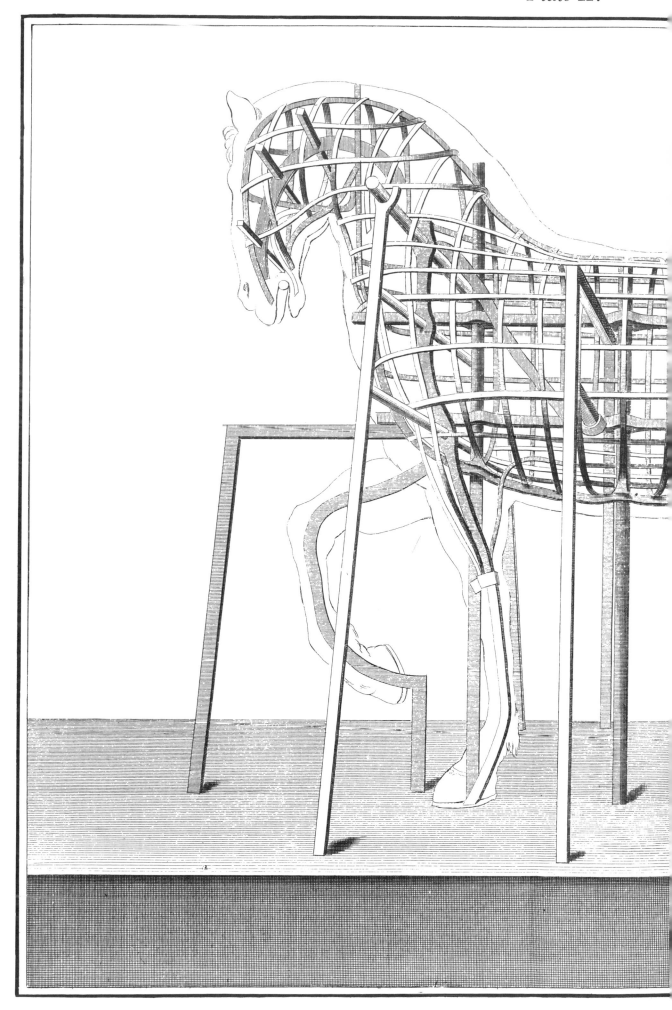

1

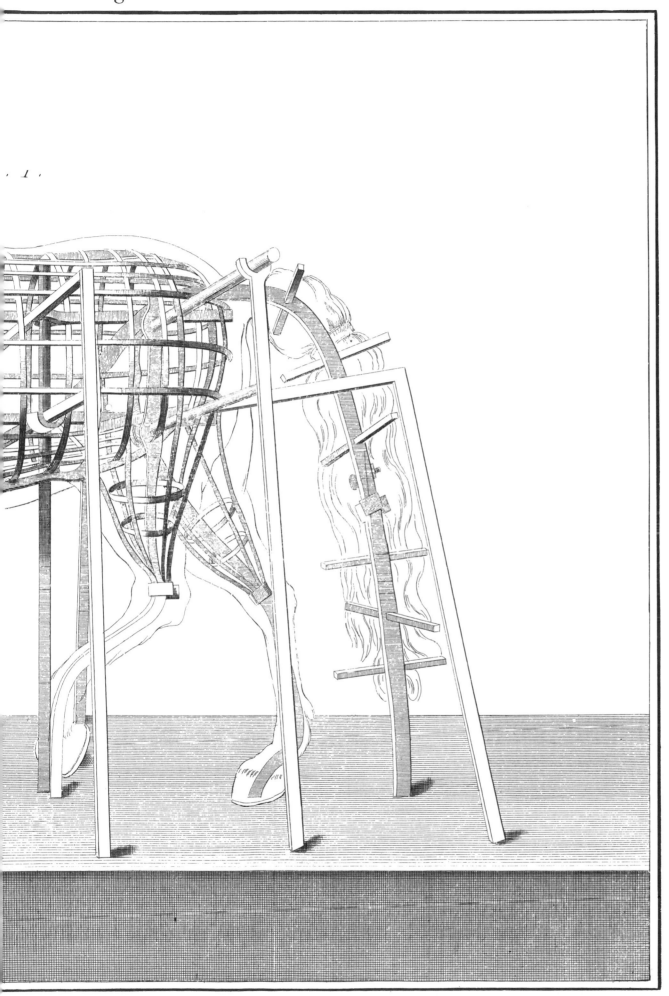

Plate 118

Fig. 2.

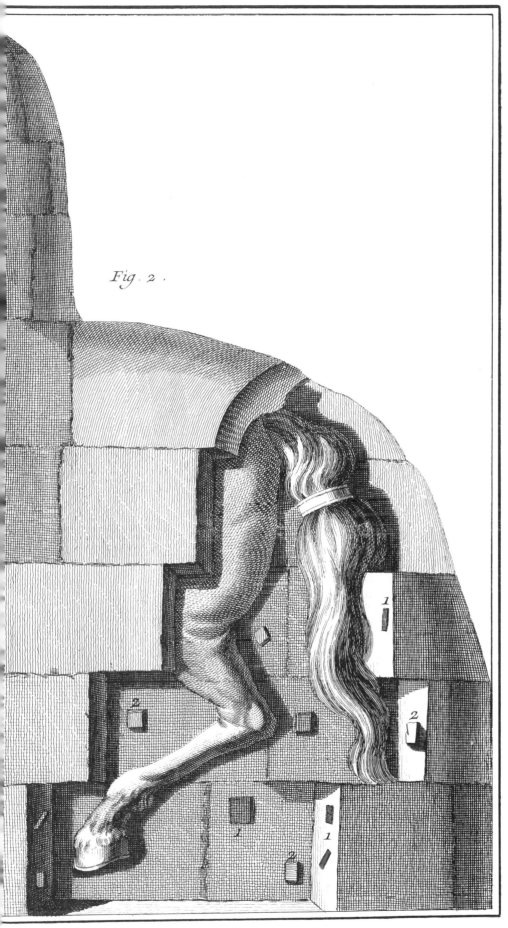

Plate 119

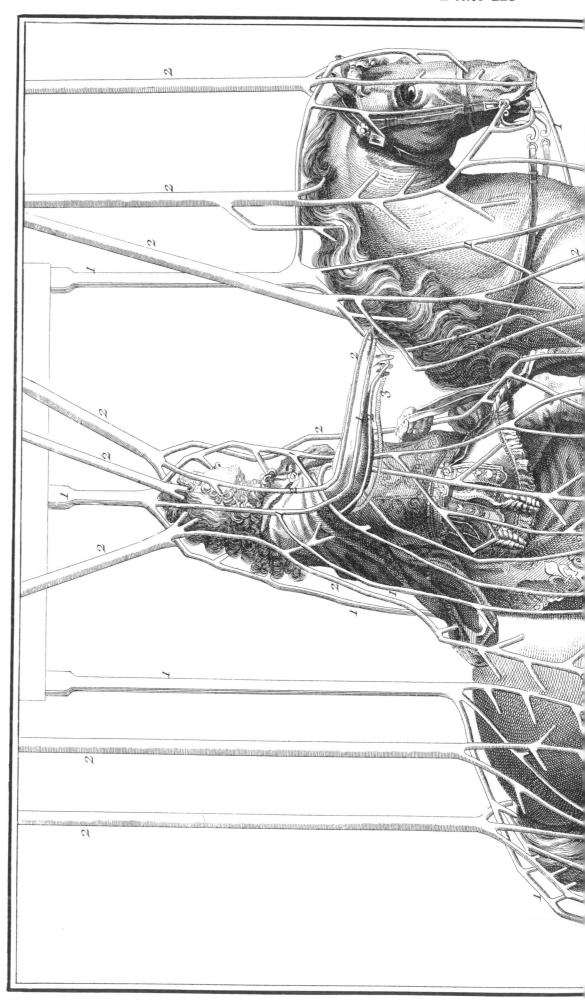

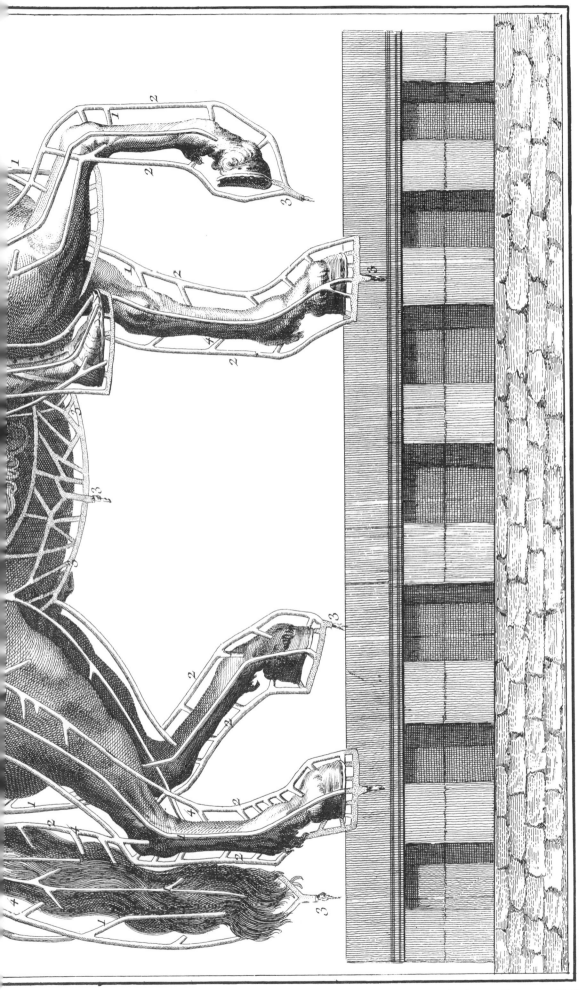

Plate 120

Fig. 1

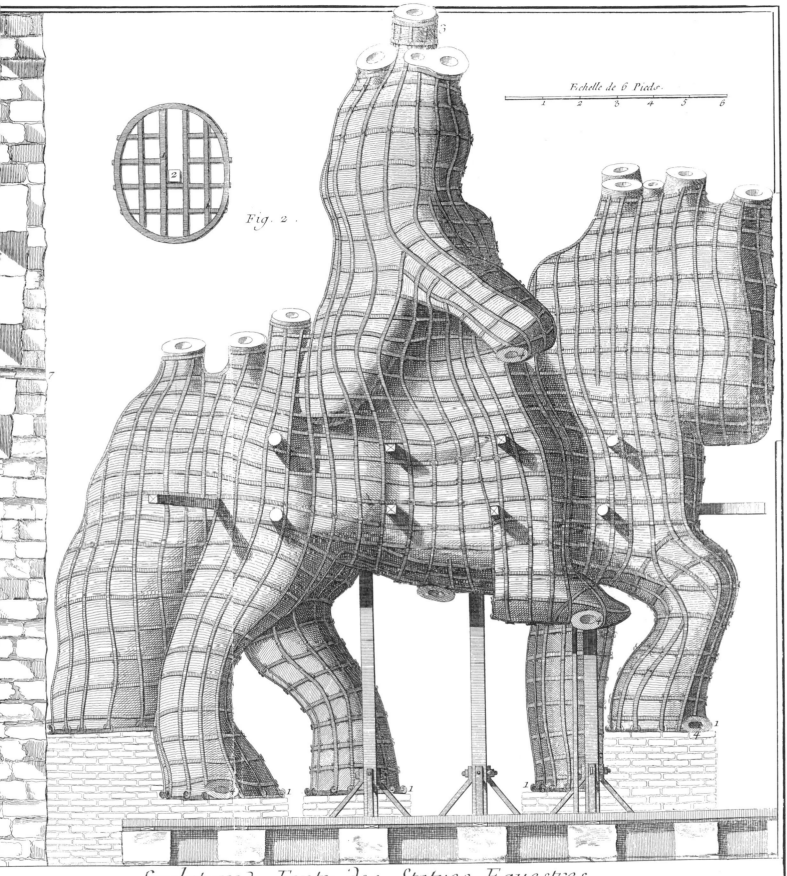

Fig. 2.

Echelle de 6 Pieds.
1   2   3   4   5   6

Sculpture, Fonte des Statues Equestres.

de la Figure Equestre par le milieu de sa longueur, avec le Noyau qui remplit la capacité renfermée par la Cire, les Egouts des Cires, les Jets, les Events, et entourée de Briquaillons. Et Figure Equestre couverte du Moule de Potée, recouvert du Bandage de fer.

Plate 121   Casting A Statue V

Extractive Industries

Extractive Industries

In the 18th century the manufacture of iron was, as it still is, the greatest of the metallurgical and extractive industries. But other ores were mined and smelted independently, and minerals which yielded commodities like salt and sulphur were widely exploited. By definition, the extractive industries draw materials from the earth, from *inside* the earth typically. Underneath the rationalism of the Enlightenment the popular feeling seems still to have lingered that mining was not, perhaps, a wholly innocent pursuit. There is something daemonic in handling metals, and Vulcan, the deformed, is the least sympathetic of gods. In folklore, the caverns of the earth are the abode of goblins, and miners are a race apart. According to old mineralogical doctrine minerals are *bred* in the earth. They grow from seeds implanted there as in a womb. These instincts were vestiges of the representations by which prehistoric man related himself to the earth mother, the source of life, and they were passed on to modern man in the collective subconscious where superstition dwells.

Superstition was the great target of scientific ideology, which so successfully attenuated these images in the 18th century that they are barely discernible. Nevertheless they did emerge, though gratefully and for the last time perhaps, in the romantic sense of nature which invested landscape with personality, and which in the plates to follow suggests all unintentionally the artist's feeling that strange places imply strange powers and that a mine was not a very wholesome place to be.

Plate 122   Vesuvius

It may originally have contributed to the dread in which mining was held that it seemed presumptuous for man to burrow towards those depths from which nature produced such manifestations of her powers as volcanic eruptions. In any case volcanos exerted a powerful fascination over the sense of the earth and her forces. Here fishermen spread their nets in the shadow of Vesuvius.

Plate 123   Volcanic Powers

Vesuvius displays her powers in the eruption of 1754.

Plate 124   Hot Sulphur

In southern Italy the road to Naples ran by the hot springs of Solfatara, where there was a perpetual bubbling and fuming of sulphurous waters. Alum was extracted from this domesticated volcano in the sheds to the left.

Plate 125   Giant's Causeway

One of the natural wonders of the world is the Giant's Causeway in Ireland's County Antrim, which the artist peoples with an excursion of fashionable nature lovers whose dress and attitudes are more French than British. In 1763 the naturalist Desmarets discovered similar structures of columnar basalt in the mountains of Auvergne—to the great satisfaction of the French—and argued, quite correctly, that these prisms were crystallized forms of lava, and that these ancient rocks had, therefore, in ages past been spewed up from the molten heart of the earth.

Plate 122

Plate 123

Plate 124

Plate 124

Plate 125

Plate 126

Plate 126   Columnar Basalt

An outcropping of the basalt of Auvergne in the Massif Central, the geologically ancient heart of France where the conical "puys" are the peaks of long extinct volcanos.

Vol. VI, Minéralogie, 6me Collection, Pl. VII.

On top of the butte are the remnants of a medieval fortress, the Chateau de la Tour d'Auvergne, and at its base is a naturally paved plaza which the village used for fairs.

Plate 127 Mining I

Mining was no new art in the 18th century, though a notable evolution of scale and technique will be apparent if the plates from the Encyclopedia *are compared to Agricola's 16th century treatise,* De Re Metallica, *translated into English by former President and Mrs. Herbert Hoover.*

Here is a general view of the essential elements of a mine, intended less to illustrate the details of operation than to evoke the atmosphere above and below ground. The artist gives us stunted and gnarled trees, not only to convey romantic and slightly sinister overtones, but to illustrate the kind of country in which metallic veins are likely to be found.

Sparse vegetation and precipitous terrain were two of the signs that encouraged the prospector. He would pay attention, too, to the sand of stream beds, to the color of soil, and to drainage. It was hopeful if rain ran right off and snow melted rapidly— this might betray the drying seepage of mineral "exhalations." But all this was very uncertain and unsatisfactory, and the best (if not the most scientific) way to find ore deposits was to come upon ruins or remains of medieval mining operations: the men of that time were more likely to have met with some non-mineralogical disaster than to have exhausted a vein.

Plate 127

Plates 128, 129 Mining II & III

There are in the Encyclopedia *a number of illustrations which do not support the text, but which were attractive enough that Diderot was, presumably, reluctant to omit them. Something of the sort might be the explanation of the device being used to locate the metallic veins in this and the next plate. It looks very like a divining rod—the "dowsing" rod of American water diviners. The text of the article on mining, however, is very severe about the* baguette divinatoire—"*on which people in some countries were weak enough to depend in seeking metals. This is a* superstitious practice, *which sound science has long since discarded." Sound science, indeed, classified the planning and design of mine shafts as "subterranean geometry," and the Encyclopedist would, presumably, have approved the rational patterning of excavation being traced out, of which the basket of ore brought up in the third plate is the due reward.*

Mining differed from other ventures requiring capital risk in that the entrepreneur was usually a landowner or nobleman rather than a bourgeois. Ownership of land as a rule carried with it mineral rights, and exploiting them did not involve the stigma attaching to commerce any more than did farming the surface.

Like everything else of which they disapproved in France, the Encyclopedists blamed the backwardness of mining on superstition and idle nobility, the one abetting the other. Diderot cited as a worthier example the energy of the German princelings, many of whom both governed and mined their little states. These rulers even maintained schools of rational mineralogy like the famous Bergakademie at Freiberg-im-Sachsen, which brought science to bear on technology. "Subterranean geometry," wrote the Encyclopedist, flourishes only in Germany, where men "have interests to discuss in the entrails of the earth."

Plate 128   Mining II

Fig. 1

Plate 129 Mining III

Fig. 1.

Fig. 2.

Plate 130 Mining IV

Fig. 2.

Plate 130 Mining IV

The method of shoring up the galleries was adapted to the inclination of the veins.

Plate 131

Fig. 1.

Plate 131   Mining V

Miners encountered many dangers and discomforts. Of all the problems, drainage and ventilation were most intractable. Fig. 2, facing, shows how the mine-shaft would, as it was sunk, encounter springs and pockets of subsurface waters (D, D, D). The water had to be carried off, either by cumbersome pumping arrangements or (preferably) by running where possible a drainage gallery out of the hillside.

The profile of a mine in Fig. 1 conveys a realistic impression of mining conditions: the darkness, the cramped spaces, the awkward digging, the crude hoists and tools, the dim torches using up what air there was. Sometimes the air could be improved by venting the mine at a different level so that it could act as a siphon, but this was never adequate.

Fig. 2.

What with the bad air and the constant fear of explosion, cave-in, or suffocation, miners were subject to various vocational hallucinations. In the most common of them there would appear hanging in the upper reaches of a gallery what the French called a ballon *— an object like a great soccer ball, covered with spiderwebs. English miners believed that this was an agglomeration of sweat from their own bodies, fumes from the lamps, and exhalation of the minerals threatened by disturbance. But it was known to miners all over Europe that should the* ballon *break, its contents would diffuse throughout the mine, killing everyone it touched.*

Plate 132   Mining VI

Fig. 2.

Plate 132   Mining VI

Miners were torn between the temptation to ease their work, by blasting, for example, and the risk of doing so. Sometimes they would build a subterranean bonfire at the end of a gallery in hope that the heat would loosen the rocks—but not too much.

Plate 133   Mining VII

Diderot appears to have turned to a less impressionable artist to illustrate the uses of mining machinery in this rather antiseptic model mine.

Plate 134   Mining VIII

This picture too is engraved in the spirit of ideal method rather than untidy reality. It illustrates the construction of various types of support, of wood or even of masonry, and the arrangement of the central shaft, with ladders for the miners and a larger area (G) in which the hoist is raised and lowered by the horse-powered winch above ground (Fig. 1). The device at the right (Fig. 2) is an ore crusher.

Plate 133   Mining VII

Plate 134   Mining VIII

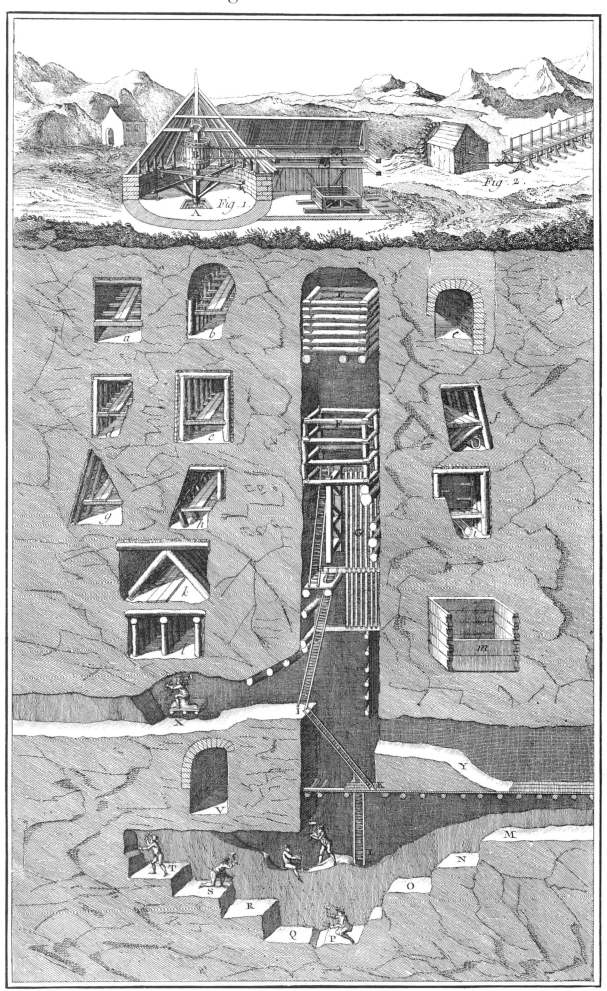

Plate 135   Salt Mining

This unusual plate depicts the most famous salt mines in Europe which underlay the little Polish village of Wielicska, near Cracow. They had been worked for centuries, and showed no sign of becoming exhausted. As will be apparent, the operation was large. There is a stable—notice the horses being lowered or hoisted in slings at the right —warehouses and packing plants, even a chapel. In the 18th century, the cavern had been dug out to about a mile in length and half a mile in width. The deeper galleries reached down about 1,000 feet beneath the level of the village which perched on the crest of this enormous excavation.

Plate 135

Plate 136   Mercury

Mercury—quicksilver, that is to say "live" silver—is the most provocative of metals in its properties. Since it is also one of the easiest to extract from its ore, it always figured prominently in the history of chemistry, alchemy, and medicine. Alchemists saw it as the "principle" of metals. Itself purer or finer than metals in its liquid state of freedom, it could infuse some quality of nobility into base substances to raise them towards the metallic. The alchemist judged matter in this way instead of simply describing its behavior. He sought the "secret" of mercury or gold, and this is what marked him off from the chemist who seeks to dissipate the unknown, not to penetrate it.

Alchemy was no longer respectable in Encyclopedic times. These two vignettes illustrate the extent of technological progress from the 16th to the 18th centuries. The upper plate is redrawn from Agricola's De Re Metallica *(p. 427), and Diderot's artist has reclothed the workmen in 18th century dress. Mercury is distilled out of its sulfide (cinnabar) in crude iron retorts (F) mounted in receivers (G). The workman is tamping a wad of moss into the retort. This will filter the metallic mercury as it condenses in the retort and trickles down into the receiver (see the arrangement F/G in the foreground). Fig. 2 shows the process of distillation carried out in a crude open hearth. The fire the artisan is poking is built around a number of retorts, arranged as in Fig. 3.*

Distillation was still the method employed in the furnace below, invented in Spanish America in the 17th century. It was in use by this time at the two chief centers of mercury production in Europe, Almaden in Spain and Idria in Austria. Both localities were rich in cinnabar. This is in essence a large distillery. The right-hand wing (A) houses two domed (CC) reverberatory furnaces. The left (K) acts as receiver. Figure 7 gives another view. The condenser is a chain of bottles (F, H),—"aludels"—joined together head to tail. Only one chain is shown, but there could be a dozen in actuality. Fig. 6 gives a cross-section of the furnace with its charge (D) of cinnabar and shows the arrangement of the aludel at the vent.

So poisonous were the fumes from such a furnace, that workers were employed only one month a year. There was no vegetation near Idria, beasts refused to eat the hay from the surrounding area, and calves were stunted in growth.

In some places mercury occurred in the free state instead of as cinnabar. According to the Encyclopedia, *that deflator of tall tales, miners who worked these rare deposits were so impregnated by the metal that a piece of gold placed in their mouths would turn white in amalgamation with the mercury in their systems. But if they were rotated to rock-crushing, the fresh air restored them.*

Plate 136   Mercury

Plate 137   Gold I

Plate 137   Gold I

Gold occurs in the free state. But it was usually so mingled with quartz and other rocks that separating it mechanically would have been impractical. In this mill, the mine products are first washed (Fig. 2), and then pulverized and mixed with mercury under the stamp (Fig. 1). Mercury forms an amalgam with gold to separate it chemically from the rock.

Plate 138   Gold II

Vol. VI, Minéralogie, et Métallurgie, Or, Pl. III.

Plate 138   Gold II

*Gold is retrieved from the amalgam (Fig. 1) by distilling out the mercury (Fig. 2) after which the gold is melted and skimmed in the crucible resting on the ovens of **Fig 4.***

Plate 139 Copper I

In the 18th century, bronze, an alloy of copper and tin, still found use in ornamental work, bell metal, and ordnance, and copper held second place among industrial metals. As in most branches of non-ferrous metallurgy, Germany was the foremost producer. The furnaces in this plate were the most modern in Saxony.

Copper was won from its ore by smelting operations which seem rather indecisive when compared to the way in which great pigs of iron were regularly born out of the belly of the blast furnace. The copper ore being processed here has been roasted to drive off arsenical impurities. The smelting furnaces (Fig. 1, B, B) are blown by water-powered bellows. Smelting a single charge might take four or five heats. Ore, charcoal, and slag from the previous run were charged as shown in the right-hand oven. The arrangement in pairs permitted continuous operation. The first metal to run into the receiver below was heavy with slag, which was lifted off as it hardened, to give way to a coarse black regulus—"matte"—of impure copper. As in successive operations, the matte was withdrawn by a spatula in the form of flat "cakes" of metal. (The left-hand oven, D, of Fig. 1 is for use with silver-bearing ores, as will be explained in Plate 141).

The firing furnace below had first to be heated and sweetened by lining with charcoal and siliceous earth. Then the cakes of matte—200 to 300 of them—were packed in layers of charcoal, and reduced to a melt in a heat of two to three hours. The foreman sampled the melt by gathering droplets on an iron rod which he plunged into a water bucket in order to observe the color of the suddenly congealed copper. He had to be very attentive, for copper reached the proper state quite suddenly, and when the moment came, the heat had to be stopped and the simmering copper left to cool in the furnace trough. Now comes the art. Copper is solidified from the surface down by successive sprinklings with cold water. If this began too soon, there would be an explosion. But before the sea-green melt had cooled far, the foreman cast across it the contents of a flask of water. This hissed into steam and left a cake of "rosette" copper to be lifted from the furnace with the spatula. The process was repeated, and the red copper cakes stacked aside, until there was left at the bottom only a lump of copper known in the industry as le roi—*the king.*

Plate 139   Copper I

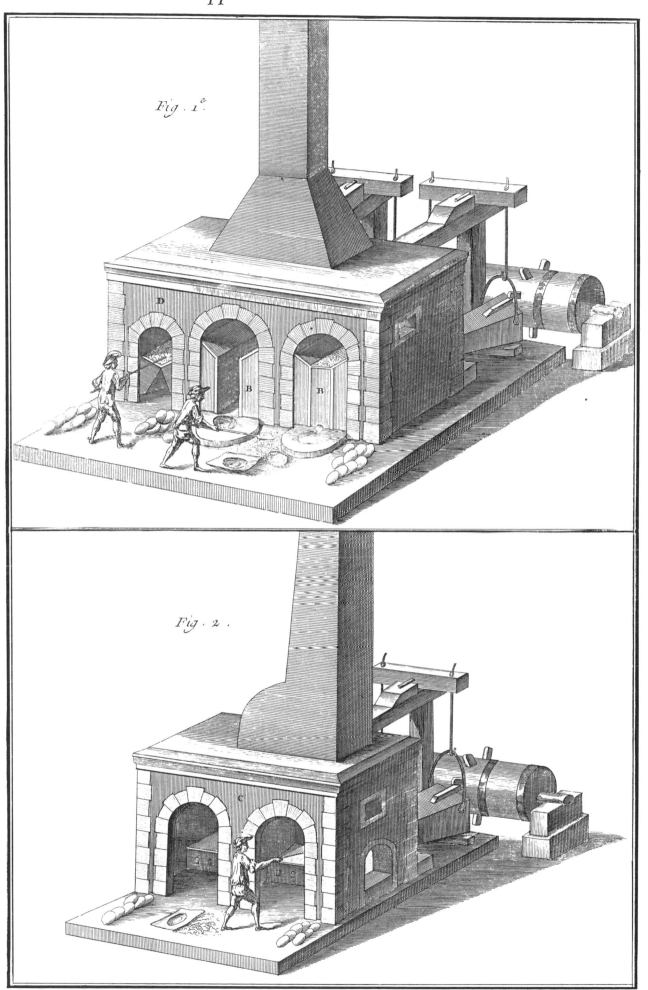

Fig. 1

Fig. 2

Plate 140   Copper II

In England copper founders preferred a reverberatory furnace of the types shown in Figs. 1 and 2. Such furnaces reflect heat from the roof down upon the material to be melted. The men ladle molten copper into shallow molds in the sand of the foundry floor, where their cakes solidify in laminated form.

So strong was technological tradition that England and Germany remained faithful to their different processes until recent times. This unreasoning variety aroused the impatience of the Encyclopedist: "Among all workmen, none are more stubbornly attached to their old methods than those who work ores, because there are none whose methods are less enlightened."

Nevertheless, this variety of methods was imposed on the copper industry, not just by benightedness, but by the differing character of the ores. Frequently, for example, copper would occur in lead and silver bearing lodes, in which case the silver would have to be reclaimed, usually by taking advantage of its solubility in lead. This is the purpose of the "liquation" furnace "D" in the previous plate (Fig. 1) and Fig. 3, G of this plate. In the former, cakes of matte are "liquated," to melt out the lead-silver alloy. In this plate, the process is being repeated at higher temperature to extract residual silver and lead from the black copper cakes on the hearth. The silver and lead are then themselves separated in the cupelling furnace of the next plate.

Plate 140   Copper II

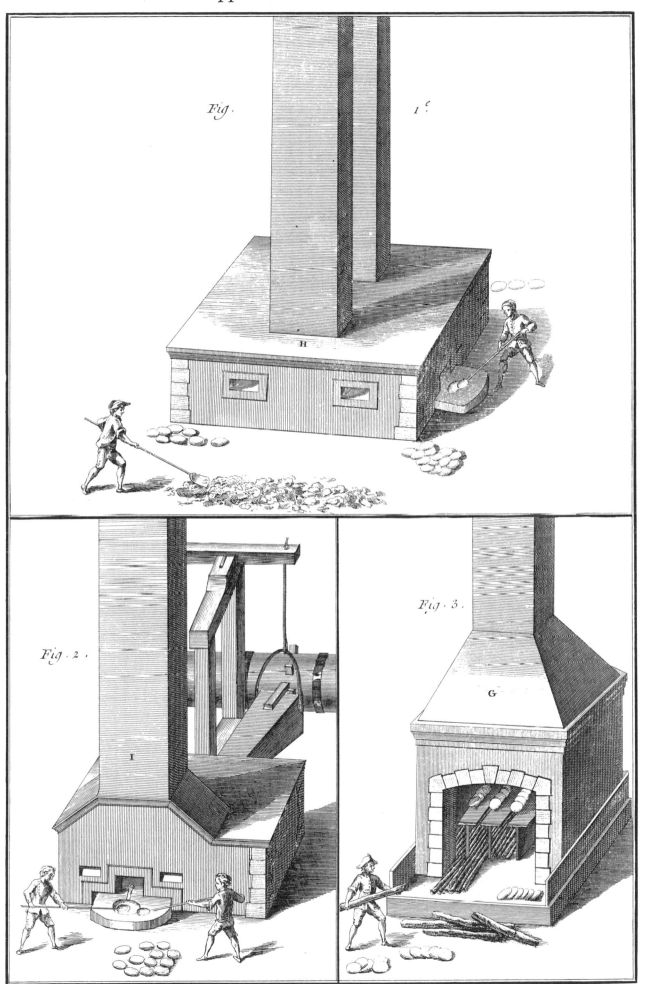

Fig. 1.

Fig. 2.

Fig. 3.

Plate 141

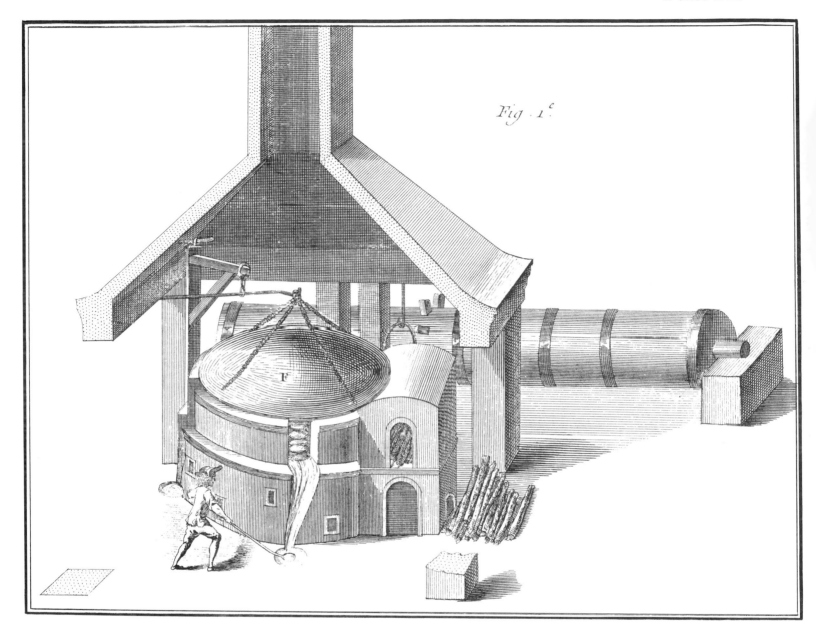

Fig. 1.

Plate 141   Silver

Cupellation was an ancient and efficient method of separating precious metals from base.

The furnace, the walls of which are cut away in Fig. 2, is blown by the inevitable water-powered bellows, and lined with a bed of wood and bone ashes, which act as a flux. The air blast oxidizes lead to litharge, which floats to the surface of the melt to be drawn off (Fig. 1), leaving the silver pure. Litharge itself could be used for pigment, or as a source of metallic lead.

Fig. 2.

Plate 142   Brass   I

This is the ensemble of a brass foundry. The alloy of copper and zinc, brass is for most purposes superior to bronze. It is harder and has greater tensile strength. It can be cast and rolled into plates. It can be drawn into wire or worked artistically. From the 13th century on, monumental brasses on important tombs began to supplant the stone effigies of the earlier Middle Ages.

Nevertheless, until a few years before the Encyclopedia *was published, the composition of brass remained a mystery. For zinc is a metallurgical fugitive: the metal sublimates at a temperature lower than that required to smelt the ore. No sooner is it liberated than it escapes as vapor.*

Brass was made, as it is here, by cementation of "rosette" copper with calamine, the zinc ore. Until the discovery of zinc itself, it seemed mysterious that mixing a metal and an "earth" (an "anti-metal" in the old chemistry), should produce a new metal of superior quality. Brass could seem a step from copper toward gold and this was often taken as an indication that the alchemists might have been on the right track, and that in zinc lay a hopeful clue. There are always temperaments that yearn after the occult, but the Encyclopedic spirit would give them no quarter, and the articles on brass conclude by pointing the moral of the discovery of zinc: "Thus, the marvels which the ignorant see in the union of calamine to red copper, and the high hopes which the alchemists found on zinc, vanish before the eyes of an enlightened man."

The foundry process was simple and continuous. The ore, smithsonite or hemimorphite, was first burned to remove impurities. A loose pile of mineral and charcoal was built, and the fire allowed to smoulder for 8 to 12 hours. This reduced the ore to a dried-out residue (Fig. 4). The ore, with loaves of rosette copper and fragments of scrap brass, was then charged into melting crucibles, which were fired in the battery of three furnaces (A, B, C, Fig. 7). The pits G and H received the waste and slag skimmed off during the melt. Each furnace held eight crucibles.

After 12 hours the molten brass was transferred into pouring crucibles (Fig. 9). It was then cast upon the rectangular pouring tables (I, K, L) in the background. Each table took the contents of a single furnace. The surfaces were made of slate and coated with a thin layer of clay.

After pouring the brass, the workman lowered the heavy lid onto the surface of the metal. This produced a uniform sheet of brass, which was cut to size by the shears shown in Figure 10.

Three founders are shown. The plate also shows the crankman, who like other manpower, would be brought in temporarily from another shop. The founders customarily ran two melts a day, from Monday through Saturday. Since the various processes required regular attention, the founders could not leave the shop, where they would take turns sleeping and where their wives would bring their meals. On Saturday night, they banked the furnaces, and spent Sunday at their cottages, which were furnished by the proprietor along with the coal, wood, and beer for each household.

Plate 142   Brass I

Plate 143

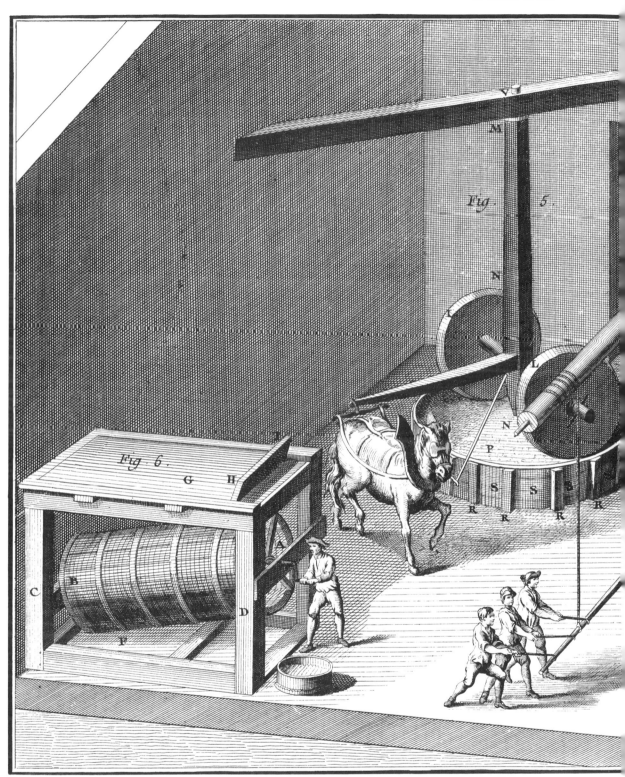

Plate 143   Brass II

Various subsidiary operations were performed in shops adjoining the foundry. Raw ore is crushed (Fig. 5), and mixed (Fig. 6). After a number of runs, the clay coating

Fig. 11.

A

Fig. 12.

c

K

d

b

b

b

of the pouring table has to be scraped off in order to be replaced. This is accomplished (Fig. 11) by spreading copper filings between the surfaces and rubbing them over each other with the apparatus shown. Finally, the leverage which worked the giant shears appears in Fig. 12.

Plate 144

Plate 144   Brass III

This shop produces hammered brass bowls, vases, and other vessels (Fig. A) from the
sheets that come from the foundry. The two batteries of hammers are powered like

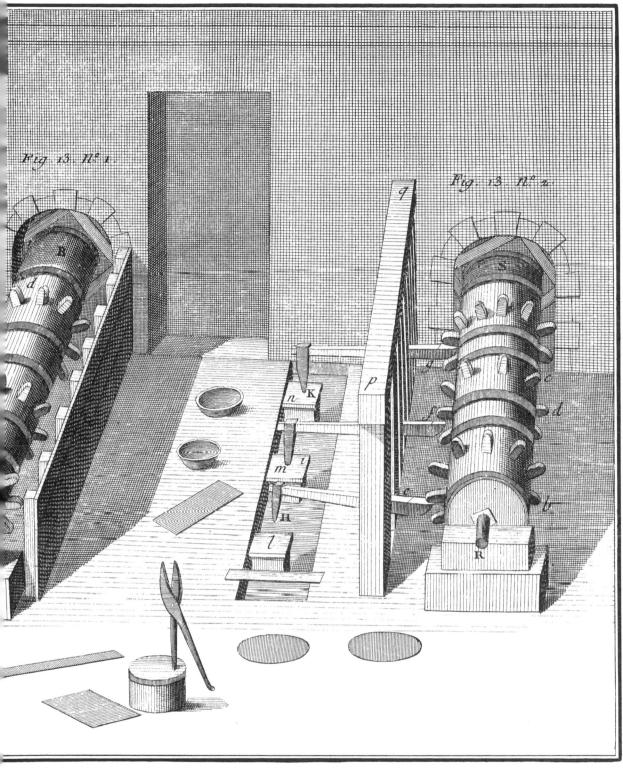

Fig. 13. N° 1.

Fig. 13. N° 2.

the larger trip hammers of the iron foundry. And the arrangement in series permits a primitive anticipation of the assembly line, or the machine shop with its multiple take-offs from a single power-source.

Plate 145

Fig. 19.

Plate 145   Brass IV

Here is an even more elaborate shop, of which the industry was very proud, for draw-ing out brass wire. The workman in Fig. 18 cuts a length from a brass rod fashioned in the forge. This rod will be drawn out by passing it through successively smaller

Fig . 18 .

holes in the machines of Fig. 19. The ground-level shop contains the shaft of a water-wheel which drives the machinery, the operation of which will be clear from the next plate.

Plate 146   Brass V

In Fig. 20 the spokes of the shaft (a) depress the elbow pieces (g) and pull the wire through the hole. When the spoke escapes, the spring (f) pulls the apparatus back into position for the next stroke. This action also opens and closes pincers (Fig. 21) which take a new bite on the wire with each stroke.

Plate 146   Brass V

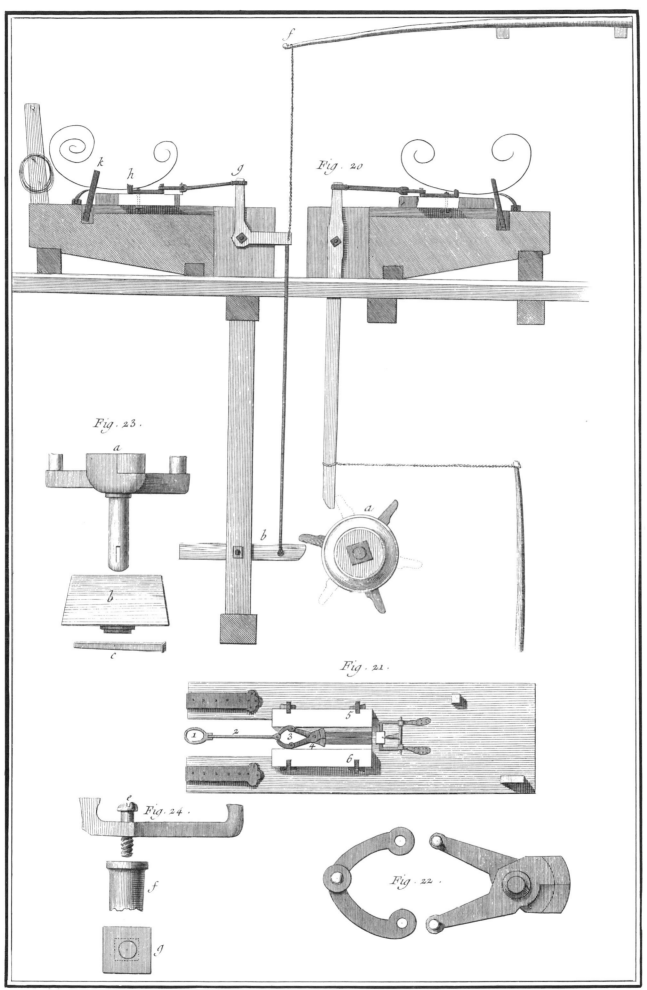

Plate 147

Fig.

Fig. 4.

Fig. 6.

Plate 147   Tinning I

*The next three plates show the process from which has evolved that prop of modern
civilization, the tin can. Plating sheet iron with tin to render it rust-proof was already
an old process in the 18th century. It had been introduced into France from Bohemia
and Germany by Colbert a little less than a century before the* Encyclopedia *was pub-*

lished, but it never flourished very vigorously and the factory shown here was located in Alsace on the Germanic fringe of eastern France.

This shop is a foundry where crude iron is melted in a furnace (Figs. 1, 6), which produces small pigs to be fired into bar iron (Fig. 4).

Plate 148

Plate 148   Tinning II

Next, bar iron is alternately forged (A) and hammered (B, C) into sheets (d, d, d) to be plated. To achieve uniformly thin sheets was not easy. The strips (B) went through three violent heats and forgings before doubling into sheets. Then they were tempered, heat-scaked in racks in a soaking oven (see inset), and the sheets of each rack

Vol. VI, Métallurgie, Fer Blanc, Pl. I.

bundled into packages for three more heats and poundings so that the center sheets could press each other into uniformity. Any that were imperfect were used for scrap. Those that passed inspection went on for tinning.

Plate 149

Plate 149   Tinning III

Tinning took many hands and many steps, but the principle was simplicity itself: sheet iron was cleaned and dipped in molten tin. But the surfaces had to be absolutely clean. Each sheet was soaked and turned from side to side in an acid solution (C) for three days. Once the surfaces were clean, they were sand-scoured and wiped by women scrubbers (G), and kept under water to prevent rusting until the tinner was ready.

The day's operations started at ten o'clock the previous evening, when the tinner lit his furnace (E) and melted his tin (F). It was essential that the tin should have been molten for at least six hours before he began to process it. Then, when the factory opened at 6 A.M., the tinner added (and here we are back in the days when each craft

Vol. VI, Métallurgie, Fer Blanc, Pl. II.

was a "mystery" reserved to the initiate) his secret ingredient, just a pinch of it, to make the tin stick. The Encyclopedist surmises that it might be copper, but all his efforts to find out were useless.

Next the tinner poured four inches of palm oil (preheated in the cauldron p) onto the surface of his tin to protect it from the air. Finally, he tinned his sheet iron, dipping several plates of iron at a time and turning them from side to side. The sheets are drained in racks (n) and wiped free of palm oil (o). A drawing of the melting furnace and bath appears at lower right, and at lower left one sees how it communicates with the flues of the furnace.

Plate 150   Sulphur

*Before the rise of modern chemistry, sulphur or in common language "brimstone"
was one of the substances which struck the imagination as portentous. But in contrast
to mercury, which was in some sense a principle of life or purification, sulphur seemed
sinister to the alchemists, a principle of fire, perhaps of damnation. Even the chemists
of the 18th century, who thought to turn their backs on superstition, associated sulphur with volcanos as a product of subterranean fire, of violent rumbling digestion in
the earth. Thus, Rouelle, the foremost chemistry teacher of 18th century Paris, told
students that the core of the earth was in perpetual conflagration fed by coal deposits,
liberating "phlogiston", the principle of combustion, which combines with the vitriol
of other minerals to produce sulphur.*

*In fact, sulphur occurs combined in many minerals, among which pyrites yielded most
readily to pre-industrial technology. Pyrites were burned for sulphur, which was
melted out by the heat of its own combustion and simultaneously driven off in the
fumes. This made the process a wasteful one, of course, and the problem of the sul-
phur-burner was to lose as little sulphur as possible.*

*In the arrangement at the top, pyrites (B) are burned against a wall (H) and the fumes
led over a water bucket (A) perched high at the apex of a roof serving as a baffle. The
sulphur distilled out of the ignition then condenses on the surface of the water.*

*The lower arrangements were more conventional. The pile of ore, Fig. 4, is built up
on a bed of fuel (G), and when it burns (Fig. 3) molten sulphur collects in the surface
pockets to be dipped up by a ladle. The smell can be imagined by anyone who steps
into a modern chemistry laboratory and concentrates in his imagination its characteristic odor. Several burned-out piles appear at the right, and there is a shed where the
crude sulphur is purified by sublimation.*

Plate 150 Sulphur

Plate 151 Vitriol

Vitriol is the old word for sulphates. Copper sulphate was blue vitriol. The more common ferrous sulphate was known as martial vitriol or copperas. Sulphuric acid itself was called oil of vitriol.

The first step in making copperas consisted of roasting pyrites and allowing the residue to effloresce in the air. This was then leached in great tanks (g, f) heated by proximity to the furnace (a, b, c) placed in the middle room of this workshop. The liquor was next evaporated (d) right over the fire, and run off by an oaken pipe, which is not shown, into the crystallizing tank (k), also oaken to resist corrosion.

The liquors still contained a great deal of vitriol. They were run into flat tanks (n, n) for final evaporation, and the vitriol stacked (t), barreled (u), and sold as a mordant to the dye industry.

This primitive process lay at the origin of the sulphuric acid industry, which at this time was about to enter on the enormous development that made of this versatile acid the most important of heavy chemicals in the 19th century.

Plate 151 *Vitriol*

Vol. VI, Minéralogie, Extraction du Vitriol ou Couperose.

Plate 152   Alum

Alum, too, was regarded as a sort of vitriol, and it is in fact a double salt of aluminum sulphate with one of the alkalis, usually potassium. As a mordant for vegetable dyes it was even more important economically than martial vitriol. In addition, its astringent action made it useful medically in staunching hemorrhage.

Medieval Europe depended on traders to the East for alum, but by the 16th century the industry had been established in most western countries. The plate shows an old alum works near Liége, one of the early centers of production. The mineral, alunite, was mined and piled in great heaps where it lay for two years in order, as the workers said, to "throw off its fire"—i.e. to oxidize a bit. Then it was calcined in a slow-burning pile in which any particular sample would smolder for eight or nine days (BB).

A workman (Fig. 1) keeps open vents for the fire. A second (Fig. 2) wheels the burnt alum to the leaching tanks. A third (Fig. 3) packs raw mineral to the burning on his back. Since the operation is continuous, the smoking alum pile crawls slowly backwards like a great burning slug consuming its own body. The fumes are fatal to nearby trees, but not—so it was thought—to the workers and their children who help haul alum.

The outdoor tanks of the upper picture form a series of leaching and settling basins, in which alum was soaked and suspended for several days before a given portion of liquor worked its way to the final reservoir (I) behind the evaporating shed.

The lower vignette moves us inside this little plant. First the solution was boiled (G) for 24 hours. After this it had another three days of quiet and clarification in settling basins (M) where crystals precipitate. From (M) the liquor, which still contains some alum, is drawn off by way of a larger reservoir (p) to replenish by a canal (q) the evaporators (G).

The crystallized alum from the settling (M) was purified in boiling pans (n), replenished from time to time with pure water. It was finally recrystallized in barrels (O). In essence, then, alum preparation consisted of alternate evaporations, concentrations, and crystallizations, all arranged so as not to waste the liquors used.

Plate 152   Alum

Plate 153   Bismuth

Vol. VI, Métallurgie, Fonte du Bismuth.

Plate 153   Bismuth

Here is another plate redrawn from Agricola and the 16th century, or possibly the late Middle Ages. It shows the preparation of bismuth. Like the first of the plates on distillation of mercury it suggests, by the contrast with the 18th century tools and procedures, how far technology had developed in the intervening years.

Bismuth fuses at low temperatures, and the metal occurs uncombined in certain minerals. All that has to be done is to build a fire around the ore and melt it out, either in series of troughs (Fig. 3) hollowed out of logs (Fig. 4), or more elegantly in a sort of open hearth (Figs. 1 and 2). In both cases a pool of molten bismuth collects in receivers (B, H).

Plate 154   Gunpowder I

Plate 154   Gunpowder I

No product of extractive industry concerned governments more immediately than gunpowder. Nor were they ever satisfied with its quality. It was, perhaps, the only item about which each state held that the others produced a better grade, a position which had the double advantage of excusing the failures of generals or admirals and urging upon manufacturers the patriotic duty to outdo themselves. In Paris the section of the Arsenal still preserves the memory of days when that establishment stood near the Bastille, but by the 18th century the principal state powder factory had moved south of the city a short distance to Essonnes.

Gunpowder—black powder as it is now called—is a mixture of saltpetre, sulphur, and charcoal. The supply of saltpetre was always the critical item. Saltpetre (potassium nitrate) forms naturally in humid soil containing decaying vegetable or animal matter, and under old French law the agents of the Royal Ordnance had the right to enter and search in any barnyard and to scrape cellar walls or requisition any particularly fruity-looking compost.

This practice was so extensively exercised that it produced one of the chief grievances of the peasants against the old regime. In this plate saltpetre is being dissolved out of humus in the upper tier of barrels, after which the solution in the lower ones will be purified by evaporation and recrystallization.

Plate 155   Gunpowder II

Plate 155   Gunpowder II

Refining saltpetre took many more steps than are indicated here, where one sees only the final stage. Nevertheless, the principle is simply boiling down the solution until cooling precipitates crystals of saltpetre which can be skimmed off with a strainer (Fig. 2). Regulations required three such recrystallizations in the refining of saltpetre in the royal arsenals.

Plate 156   Gunpowder III

Vol. VI, Minéralogie, Fabrique de la Poudre à Canon, Pl. IV.

Plate 156   Gunpowder III

This is the interior of one of the powder mills at Essonnes, showing two batteries of power driven mortars. Both are arranged in three sets of four mortars, which pound the charcoal, sulphur, and saltpetre into a progressively finer and more homogeneous mixture. Normally one workman tends each set of three, though for clarity only two of the attendants of the right-hand battery are shown.

The process begins with 15 pounds of saltpetre, 2-1/2 of pulverized sulphur, and 2-1/2 of powdered charcoal, moistened with a pint of water placed in each mortar. The machinery is set in motion for an hour, after which the contents of each mortar are scraped into the next—the mixture from the second to the first, the third to the second, the fourth to the third, and the first to the fourth. The process is repeated at three hour intervals and with occasional moistening until all of the powder has been beaten in each of the four mortars.

The pestles pound up and down at a rhythm of fifty-four to fifty-six strokes a minute. After 20 to 24 hours of this, the powder would usually be ready for drying and graining. Needless to say no one was astonished when one of these mills blew up. In order to channel the damage upwards, the roof planks are not nailed down: they are simply laid across the rafters, a precaution which protected only the men in the other shops.

Plate 157   Gunpowder IV

Vol. VI, Minéralogie, Fabrique de la Poudre à Canon, Pl. IX.

Plate 157   Gunpowder IV

This is a roller mill for pulverizing gunpowder.

Plate 158   Gunpowder V

Plate 158   Gunpowder V

After drying, powder is "grained," or sifted according to the size of granules. Its explosive characteristics depend in very large part on the size of the granules.

Plate 159   Slate I

European civilization is housed in stone, and quarrying was perhaps the most exten-sive, and certainly the simplest, exploitation worked on the resources of the earth. The plates that follow illustrate the mining of slate, which found use mainly in roofing—the line between slate and tile divides northern and southern Europe—but also in building blocks.

Two regions of France possessed great deposits of slate of usable consistency: the valley of the Meuse and the province of Anjou. Along the Meuse the laminated veins tilt steeply downward, and tempt the owner to push his shafts ever deeper. The quality of slate tended to improve with depth and pressure. This might require a consider-able investment before rewarding deposits were reached. In estimating the promise of an outcropping, the entrepreneur is warned by the Encyclopedia *not to trust the judg-ment of local miners, whose supposed peasant wisdom is usually ignorant superstition.*

In Fig. 1, the shaft of a mine under the village of Rimogne has reached a fine deposit of slate. The layers lie evenly. The workers can split out slate in steps as if on a great staircase. But there is nothing of that about the ladder (Fig. 2) up which the men struggle bearing slates. Each slate-block weighs about 200 pounds. In this mine the custom was for the workers below ground to carry the slates halfway up, where they transferred their burden to a worker from the slatecutter's shop who came down to meet them at midpoint.

Plate 159   Slate I

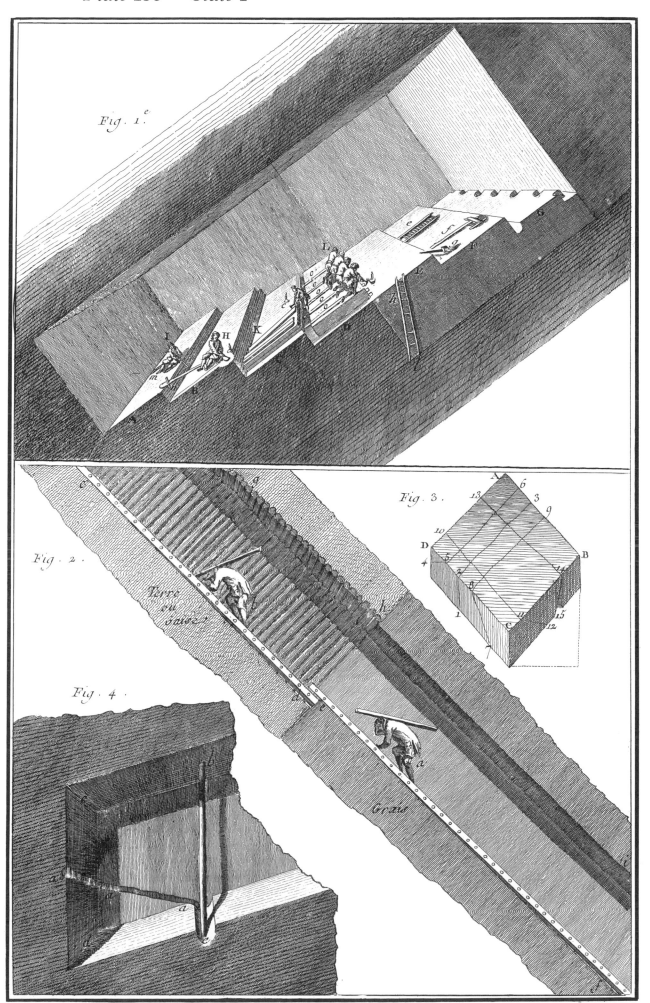

Plate 160   Slate II

The stonecutter's hut stands at the minehead. The artist has arranged to show in its roof the use of slate on a simple shelter. Neatly packed at the right (d) are bundles of slates ready for the factor who markets the product for the mine owner. A boy (c) hauls away fragments (4) to be used for fill. These have more interest than might be supposed, for the more there were of them, the less the stonecutter earned. His trade was paid in the peculiarly nerve-wracking way in which diamond cutters are recompensed. There was a premium on his getting the largest possible number of finished slates from the least amount of rock. The workers underground were paid by weight alone. The value of the slate they delivered was debited against the piece rates paid the cutter. On an unlucky day, therefore, when all his blows went wrong, he might even end in debt to the factor. With all the good fortune and skill possible, a man still had to work very fast in order to earn much, and it is no wonder that it took a long time to train one of these hacheurs.

Plate 160   Slate II

Fig. 1.

Fig. 2.

Plate 161   Slate III

Plate 161   Slate III

The slate deposits of Anjou are accessible enough to be worked in open quarries, and they yield blocks as well as slabs. Indeed, the provincial capital of Angers is built of slate which gives it, in the Encyclopedist's eye, a mournful appearance. Slabs are taken vertically. A new cut is being started at the right by means of iron wedges (K). Other tools needed are scattered about in the background.

Plate 162   Slate IV

Plate 162   Slate IV

Slitting and cutting the Angevin slates appears to be a rather more relaxed and pastoral undertaking than in the grimmer quarries of the always more industrial north. Whether the work was organized and paid in the same Stakhanovite fashion, the En-cyclopedia does not say. But it does not look so.

Plate 163   Slate V

Vol. VI, Minéralogie, Ardoises d'Anjou, Pl. II.

Plate 163   Slate V

Here is labor-saving machinery at work. The horse-powered windlass is bailing out the quarry, but it might equally well be used to hoist stone to ground level. In a time-motion study accompanying the plates, the analyst finds that it takes a horse eight minutes to lift this barrel 150 feet, and that in an hour's time the horse will have walked about two and one half miles.

Plate 164   Slate VI

Plate 164   Slate VI

The mechanical linkage in this arrangement is found to be more efficient. What is more, the horse appears to be stepping out more briskly. Something does appear to have gone wrong, but we are not told what instructions are being given to the pitman.

Metal Working

Plate 165   Ironwork I

Plates 165, 166   Ironwork I & II

The preceding two sections illustrate the winning of metals, and in general carry them through the foundry stage. After this they were in a form suited to the needs of the artisan who worked up into finished goods the things made of metal—from horseshoes, pins, needles, blades, and lead-pipe to coin of the realm.

The ornamental ironworkers of France were as superior to those of England as the forge and foundrymasters of England were to their French rivals. Indeed, the term serrurerie *has no exact counterpart in English. It embraces in a single trade fine work in iron: structural parts for architectural use, ornamental grillwork, and locksmithing. By 19th or 20th century standards, structural iron played a relatively minor part in the 18th century, whether in buildings or in ships. The time had not come for iron girders to replace wooden beams or joists or arches of masonry. But iron did serve in incidental places: in angle-irons; hinges; window-bars; downspouts; and in those alarming curved spike fences which divide even the roofs of one building from its neighbor in cities built all cheek by jowl, like 18th century Paris. The* maître-serrurier, *whose forge (above) and yard (opposite) were prepared to fill almost any architectural order.*

Plate 166   Ironwork II

Plate 167 Ironwork III

*To compare the Vieux Carré of New Orleans with other American cities is to epitom-
ize the contrast between French awareness of the decorative possibilities of iron and
Anglo-Saxon concentration on its utility. Here are various designs to be carried out
in wrought iron: fanlights (Figs. 111, 112); balconies (Figs. 113, 115); stairrails (Figs.
116, 117).*

Plate 168 Ironwork IV

*The gateway to the park of the chateau of Maisons (Fig. 129), and a portion of the
fence (Fig. 130).*

Plate 167   Ironwork III

Fig . 111 .

Fig . 113 .

Fig . 112 .

Fig . 114 .

Fig . 115 .

Fig . 117 .

Fig . 116 .

Plate 168

Fig. 130.

Fig . 129 .

Plate 169

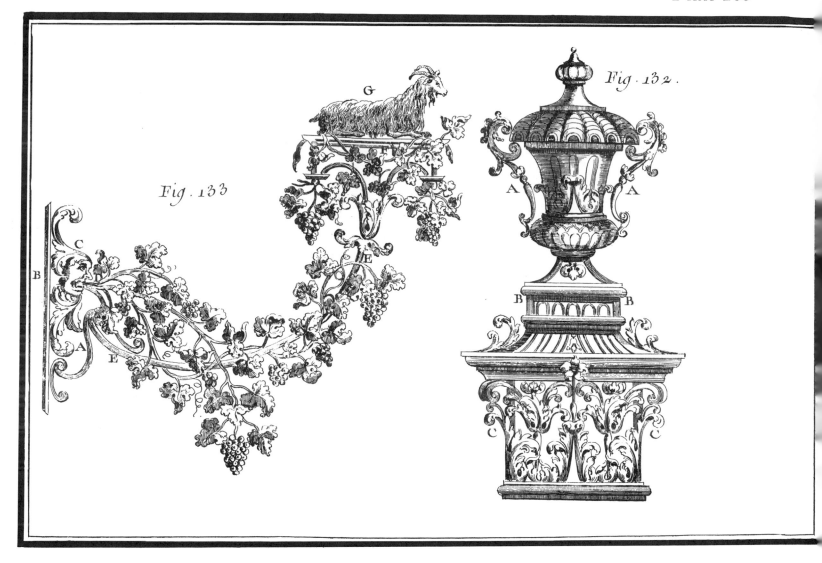

Fig. 133

Fig. 132.

Plate 169   Ironwork V

The crowning superstructure over a gateway, or a portion of it (Fig. 131); a garden vase (Fig. 132); two sign-holders for shops or hostelries (Figs. 133, 134).

Ironwork V

Fig. 134.

Fig. 131.

Plate 170

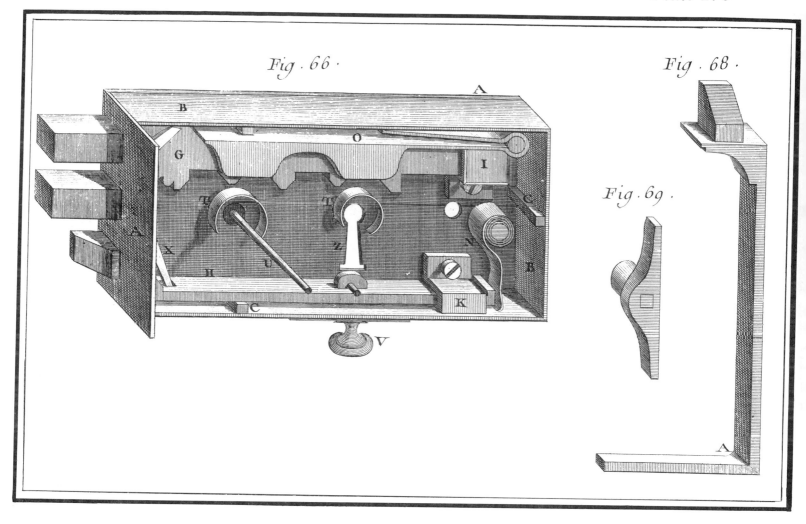

Fig. 66.

Fig. 68.

Fig. 69.

Plate 170 Ironwork VI

It is well known that Louis XVI vastly preferred his hobby of locksmithing to his duty as a king, and it was a great pity for him, and perhaps for France, that the intricacies of French politics interested him so much less than the intricacies of French locks.

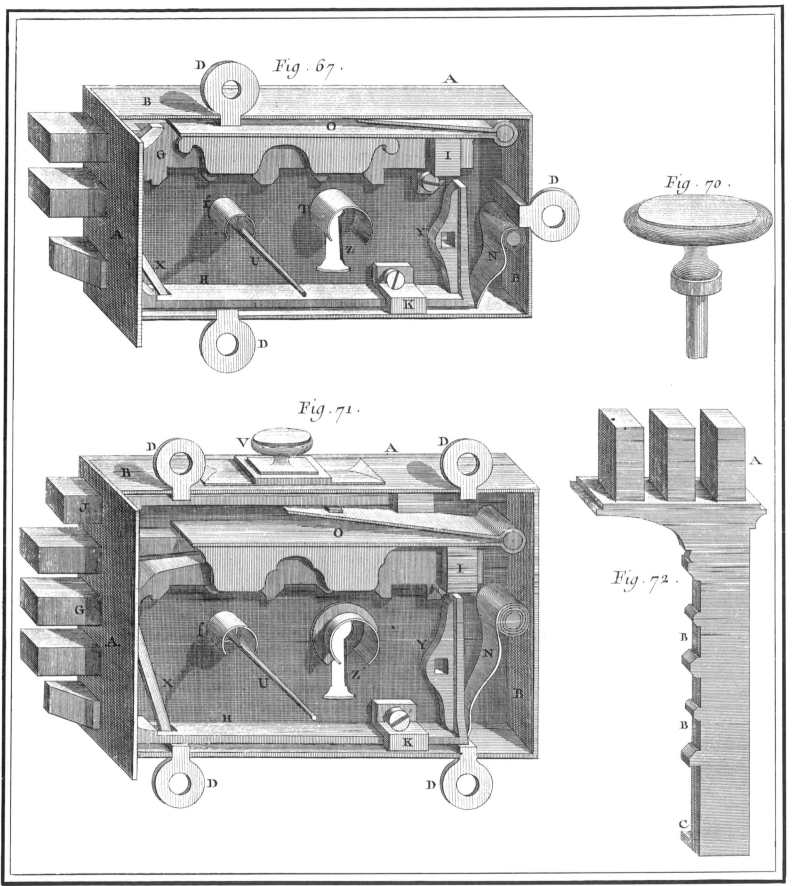

Fig. 67.

Fig. 70.

Fig. 71.

Fig. 72.

Plates 171, 172 The Toolmaker I & II

Two illustrations from the toolmaker's trade are worth reproducing. In the upper the smiths are at work on an article of prime importance to their own trade—an anvil. One man (b) turns the great block from side to side while a second holds over it a new layer of steel to be welded into one body under the hammer (d). The bellows appears to be worked by manpower—or perhaps one should say by man-weight—for apparently the bellowsman (a) compresses it by jumping on the upper side.

The lower vignette represents a much more advanced level of machining. It is a shop for threading screws and worms. In Fig. 1 the workman is tracing out the pattern of a thread on a steel rod. The machine of Fig. 3 is a threader, as on a much larger scale is that of Fig. 4. The latter is turned by the wheel (5), which is cammed to reverse the rotation while moving back and forth the width of the thread with each revolution.

Plate 173

Plate 173   Shoeing Horses I

One wonders how long it will be before this trade will have to be identified. Indeed, how many people even now know what a farrier is? In Fig. 1 a farrier plies his trade, while behind him a lackey waits his turn with his master's horse to be shod. In the background a horse is confined in a framework which makes him hold still while his hooves are pared. An enlargement of this contraption appears at the right.

Fig . 1ᵉ.

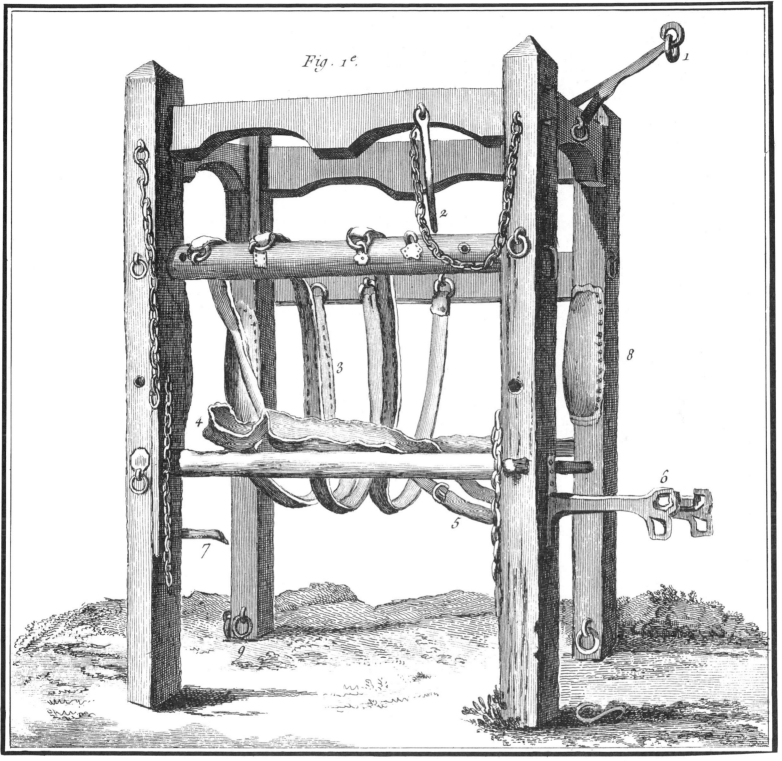

Plate 174

Plate 174 Shoeing Horses II

The farrier at his forge, surrounded by his tools and horseshoes. Notice particularly how he blows his bellows.

Plate 175   A Heavy Smithy I

Plate 175   A Heavy Smithy I

The American blacksmith turned his hand to everything, but in France occupations were more specialized. The maréchal ferrant et opérant *(the English farrier) tended horses, combining shoeing with veterinary services. The smith's heavier work was the province of the* maréchal grossier, *whose apparently frantic employees are at work on wagon wheels. The six men in the background (8, 9) are threading an axle-tree. The others are rimming a wheel. Figs. 10 and 11 show a front wheel and the iron band which will be mounted on it. The workers of Figs. 1-4 are prizing and hammering a rim over the wheel. The two men on the right are driving home nails.*

Plate 176   A Heavy Smithy II

This is the forge of the maréchal grossier.

Plate 176   A Heavy Smithy II

Plate 177 The Spurmaker I

Spurs and bits constituted the trade of the spurmaker. Designs showed great variety in detail and gave play to fancy and ingenuity, but the process of forging created no special problems. Once forged, however, a bit had to be tinplated against the corrosive effect of the horse's saliva and the wearing action of its mouth. Besides being unsightly, a rusty bit would gall and ruin a good horse.

Tinning is carried out right in the forge. The iron is oiled with a rabbit's foot (Fig. 1), and dusted with powdered wax (Fig. 2). It is then dipped in a bath of tin attached to the forge (Fig. 3). The tinner hands it, finally, to a "shaker" (Fig. 4) whose simple job it is to shake off droplets of excess tin before they harden.

Plate 177   The Spurmaker I

Plate 178

Plate 178   The Spurmaker II

A passerby looking in over the counter has this view of the spurmaker's shop. One man rivets (Fig. 1) a fastener to the corner of a mouthpiece. Beside him his companion (Fig. 2) polishes a completed bit. Back at the bellows the fire is blown up to reheat a batch of worn curb-chains which are to be retinned. They will be carried to a red heat, scaled on the anvil, dropped in cold water, and shaken in the rotating barrel of Fig. 4 to rub off the last bits of rust. Finally, they are stirred about in a cast iron cauldron (Fig. 5) containing a bath of molten tin from which they emerge as good as new—or almost.

The Spurmaker II

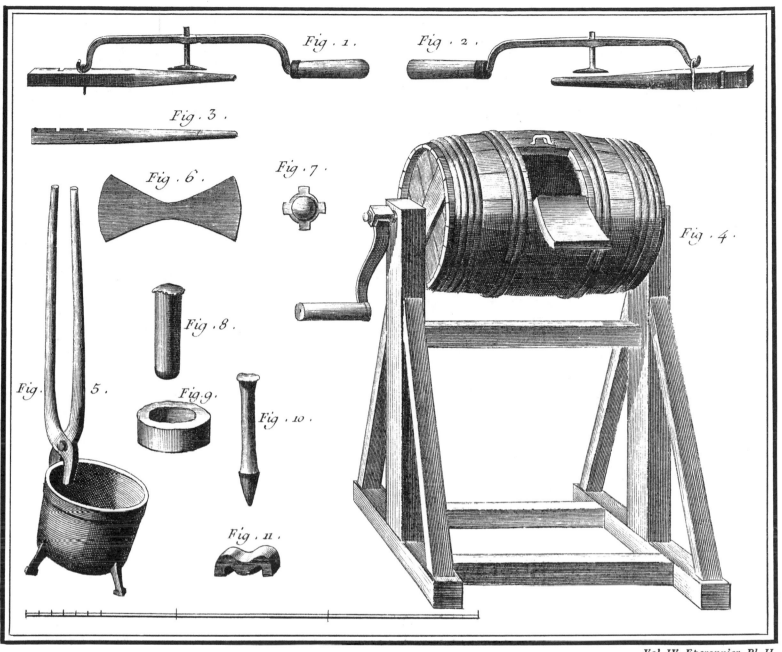

Fig . 1 .

Fig . 2 .

Fig . 3 .

Fig . 6 .

Fig . 7 .

Fig . 4 .

Fig . 8 .

Fig . 5 .

Fig . 9 .

Fig . 10 .

Fig . 11 .

Plate 179   The Cutler

Plate 179   The Cutler

Another of the Parisian shops open to the gaze of the curious is that of the cutler, whose guild had the right to forge, fabricate, and sell knives, scissors, razors, surgical instruments, and other edged implements. He bought his steel in bars, forged the blades, and ground his edges on the wheel (Fig. 2), which imposed on the grinder a posture so characteristic as to have been almost emblematic of the trade. Working at the vise are two journeymen, one (Fig. 5) with a file and the other (Fig. 4) with a bow-drill for boring. An apprentice at the desk gives a razor its final edge, and Madame, leaving the cash drawer for the moment, arranges her wares in the showcase.

Plate 180   The Tinsmith's

Plate 180   The Tinsmith's

Kitchen utensils were manufactured by the tinsmith, who displays items of tinware on his counter and hanging in the shopfront. Among his products were oil lamps of various designs. His own shop is lit by a ceiling lamp which for lighting or refilling is raised and lowered by a cord. At the moment the men are making coffeepots. The artisan in Fig. 1 fashions the body of the pot on a small anvil. Next to him (Fig. 2) the rim is filed to ensure a tight-fitting cover. At the workbench the master solders the spout into place. He has his strips of soldering tin (d) before him, and heats his irons in the brazier (b). On the floor (Fig. 6) are cutouts, one (e) for making covers (h) and the other (f) to be bent around to make the mouth of the big funnel (g).

Plate 181

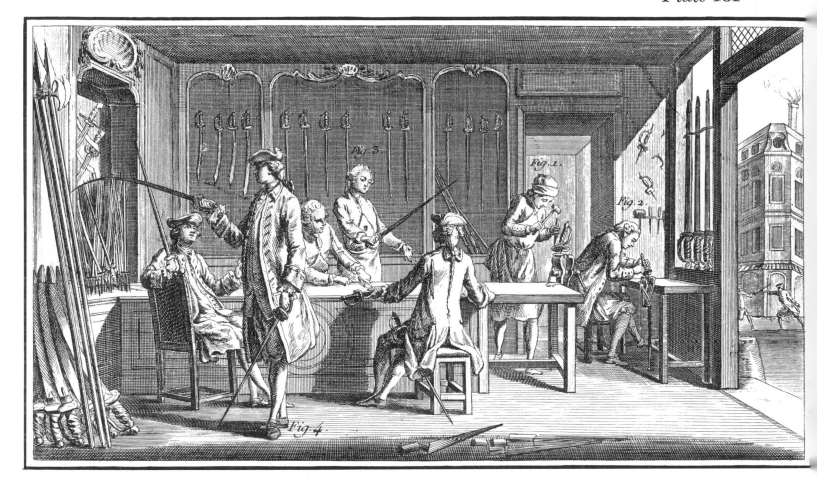

Plates 181, 182   The Swordmaker I & II

The more impractical dueling became, the more honorable it seemed, and one of the minor boons of the French Revolution was that it opened the opportunity to everybody. No longer could a nobleman like the Duc de Rohan humiliate a commoner like Voltaire by ignoring his challenge and having him beaten by lackeys. With the challenge issued in 1955 by the Minister of Finance to the publisher of L'Express, a newspaper which had insulted him, duelling under the Fourth Republic may be said to have become a part of the democratic process.

In the 18th century, however, injustice was still rampant, and dueling was the prerogative of the nobility. There is a very evident social distinction between client and craftsman in this swordmaker's shop. Besides the genteel weapons on the back wall, the proprietor also makes the halberds, pikes, and épées stacked on the wall at the left. Notice the emblem of his trade in the street window in front of the worker who is damascening a hilt (Fig. 2). Judging from the glimpse of the street, the owner need not fret about business.

Plate 182 illustrates the ornamentation of weapons.

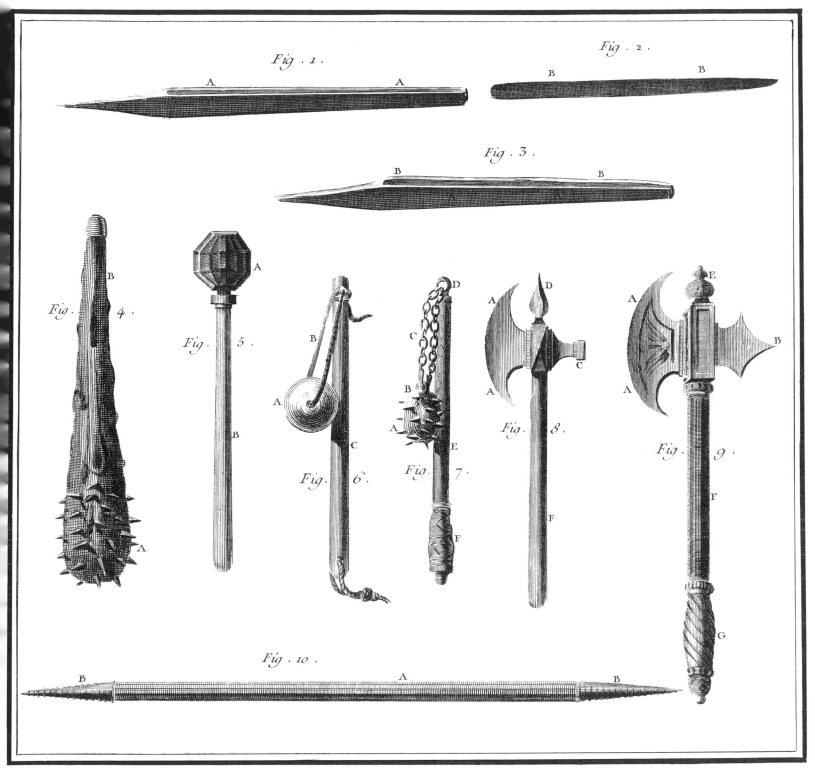

Fig . 1 .

Fig . 2 .

Fig . 3 .

Fig . 4 .

Fig . 5 .

Fig . 6 .

Fig . 7 .

Fig . 8 .

Fig . 9 .

Fig . 10 .

Plate 182   The Swordmaker II

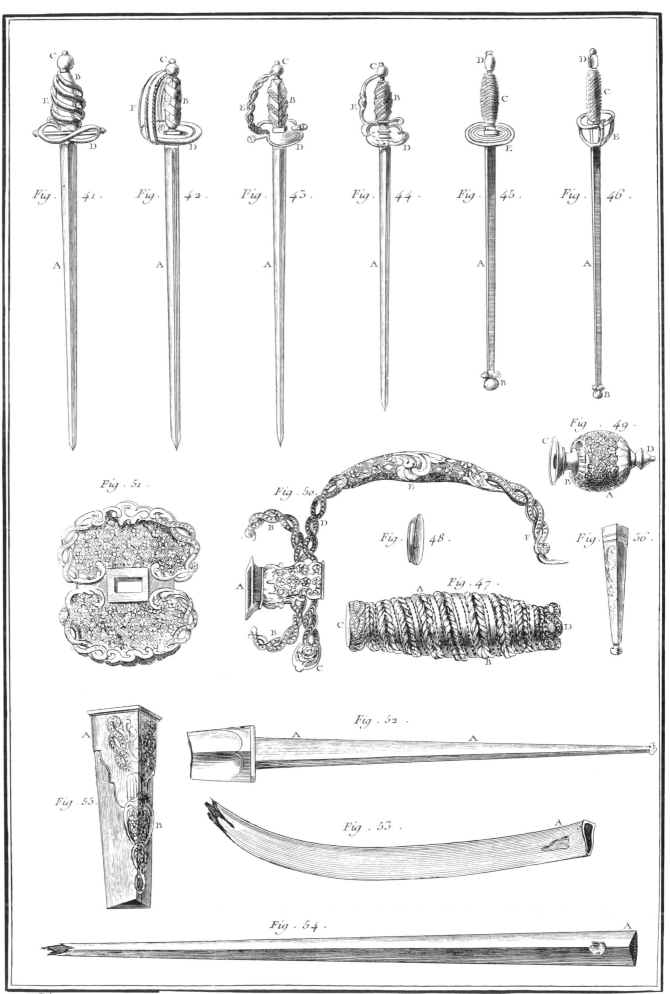

Fig. 41.   Fig. 42.   Fig. 43.   Fig. 44.   Fig. 45.   Fig. 46.

Fig. 49.

Fig. 51.   Fig. 50.   Fig. 48.   Fig. 56.

Fig. 47.

Fig. 52.

Fig. 55.   Fig. 53.

Fig. 54.

Lucotte Del

Plate 183   The Swordmaker III

Vol. IV, Fourbisseur, Pl. VII.

Plate 183   The Swordmaker III

The workmen in the shop emboss, gild, and embellish guards and hilts and affix them to the blade, but the edge must be ground by a cutler working over his wheel. Until very recently, the grinders of Thiers, the cutlery center, still assumed this posture in their trade.

Plate 184   The Pin Factory I

Plate 184   The Pin Factory I

A pin factory was the classic example adduced by Adam Smith to illustrate the advantages of a division of labor. How long, he asked rhetorically, would it take to make a pin if one workman performed every step himself?

Pins were made from wire, and since steel wire was prohibitively expensive, and iron wire insufficiently ductile, the best grade of pins was made of brass. This seems to be a rather small shop, with workmen winding (Fig. 1), unwinding (Fig. 3), and washing (Fig. 2) coils of wire before drawing it. The phrase "wire-drawn" for something pulled thin is taken from the action of this machine (Fig. 4).

Plate 185   The Pin Factory II

Drawing out brass wire (Fig. 2), although it looks the simplest of tasks, required ex-perience and a very sure touch. Everything depended on the precise alignment and spacing of the arrangement of nails (Fig. 2, d) through which the wire is pulled (see, too, Figs. 13, 17, below). The farther apart they were and the more obtuse the angle, the thicker was the wire and the heavier the pin. Each drawer made his own "engine" every time the order came down to change the characteristics of the pin and to make a different weight.

A good drawer could pull 60 feet of wire a minute off the reel, after which it passes in lengths to the cutters (Figs. 3 & 4) who snip it into pin-lengths. Their cutters appear below (Figs. 11 & 12), along with the sort of knee-vise (Fig. 21) they use to measure and hold the wire. The men were supposed to cut 70 pins a minute, 4,200 an hour. Point-ing the pins is a simple matter of grinding one end of the shank (Figs. 5 & 6, and 1, 16 below). The wheel at the right (Fig. 9) is used for heading, as will appear in the plate that follows.

Plate 186   The Pin Factory III

The central operation here is heading. Heading-wire from Fig. 9 of Plate 185 is cut into spiral coils. The butt of the pin-shank catches in the spiral, two turns of which are left around the shank, and then the head is flattened into a button by anneal-ing (Fig. 7, n) and pounding it against the miniature anvils of the machine shown as Fig. 12 below (top view, Fig. 18). This was a complicated and precise piece of machinery, a sort of tiny drop-forge.

To blanch the pins they must, finally, be tin-coated. For this they are packed in layers of tinfoil (Fig. 9) and boiled in a dilute solution of salts of tartrate (Fig. 7, m) from which they emerge to be washed (Fig. 1), dried (Fig. 2), polished by mutual friction (Fig. 5), and sorted and packed in quart jugs for shipping (Fig. 3).

Plate 185

Fig. 6.

Fig. 8.

Fig. 5.

Fig. 7.

A   Fig. 1.   B

Fig. 15.

Fig. 21.

Fig. 16.   Fig. 16. N° 2.

Fig. 21.

1   2   3
3   6   9   12   Pieds

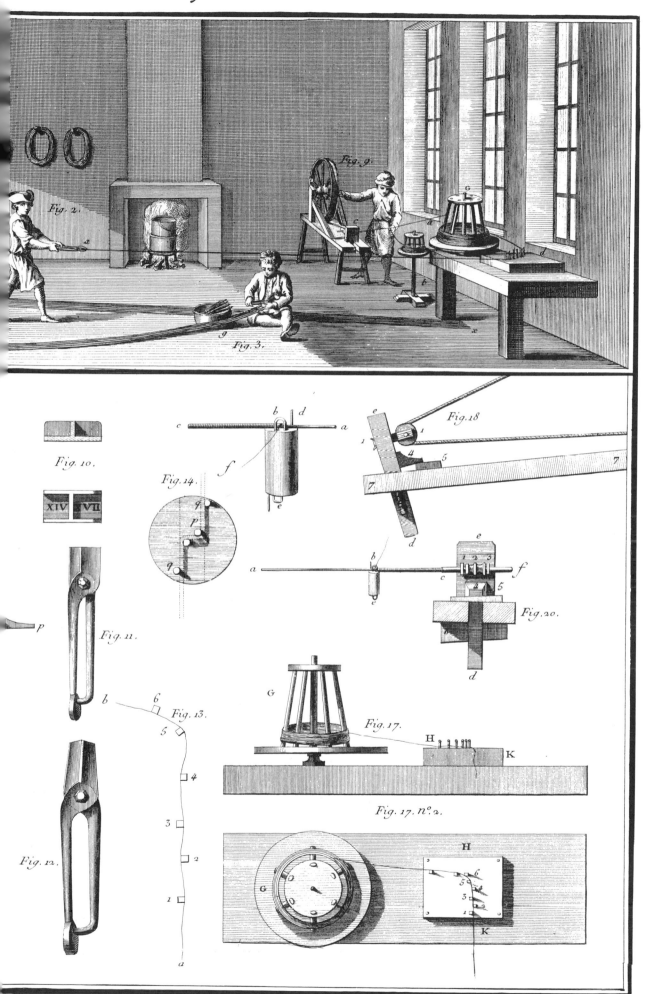

Plate 186

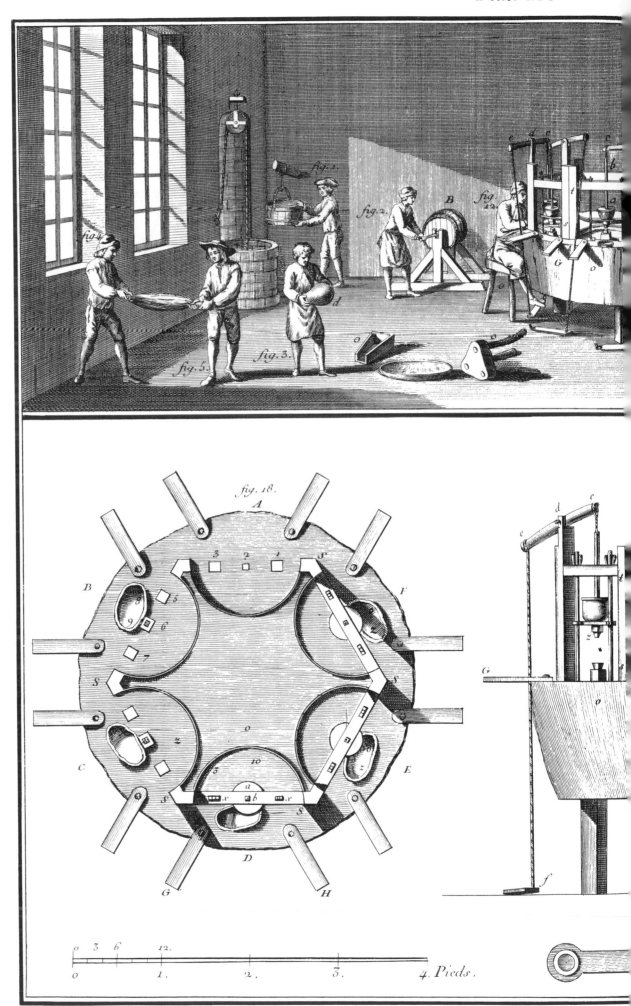

The Pin Factory III

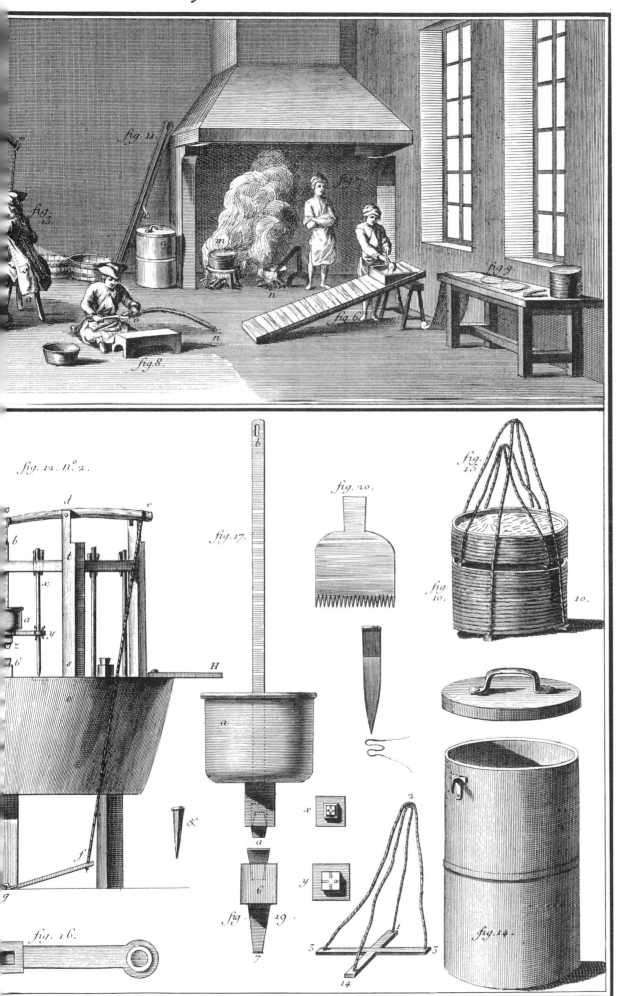

Plate 187

Plate 187   Making Needles I

Needle-making was a trade, simple in principle but skilled in practice, which might be taken as an illustration that both division of labor and a consequent monotony of routine were characteristic of certain branches of manufacture long before the Industrial Revolution. The statutes of the Parisian guild of needle-makers date from 1599, nor had technique changed appreciably since then.

Needles are cut from steel wire (Fig. 1), and the ends are then flattened, four at one blow (Fig. 4), to receive the eye. The flattened end is punctured (Fig. 2) and the resulting fragment of steel punched out (Fig. 3). After this the pointer (Fig. 7) uses a file to bevel the eye and to point the end. The next steps are tempering (Fig. 5) and reheating (Fig. 6) to allow any crooked needles to be straightened under a hammer. The last step in this plate (Fig. 8) spreads the needles, 12 or 15 thousand of them, on a piece of canvas which is sprinkled with powdered emery and a little oil and rolled up into a tight roll so that the needles may be polished by the friction of the emery.

Making Needles I

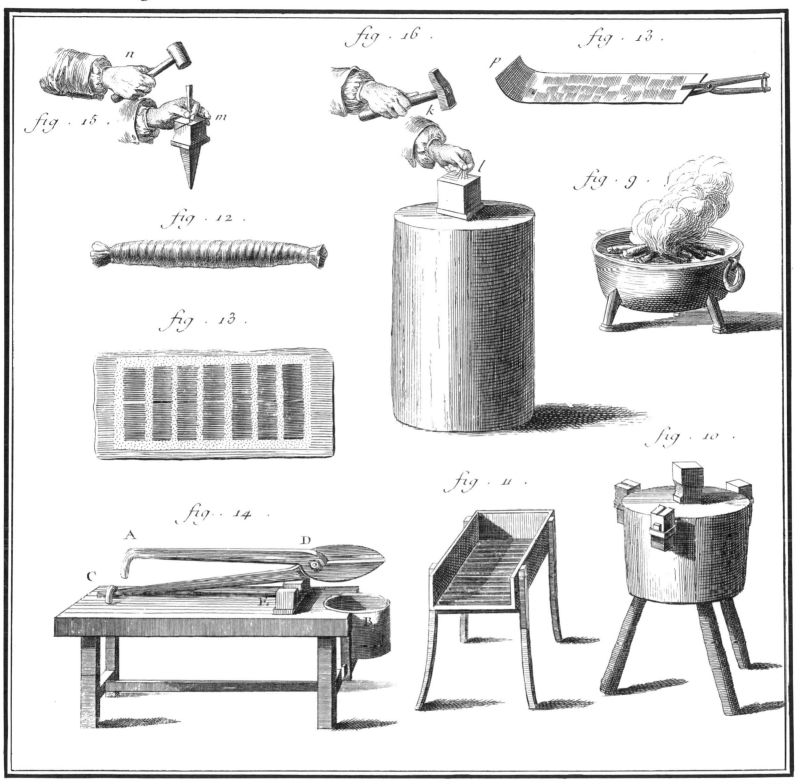

fig . 15 .

n

m

fig . 16 .

k

l

fig . 13 .

p

fig . 12 .

fig . 9 .

fig . 13 .

fig . 10 .

fig . 14 .

fig . 11 .

A

D

C

E

B

Plate 188   Making Needles II

Plate 188   Making Needles II

A workman ties up the polishing-roll (Fig. 1). At the right two men (Figs. 5 and 6) ro-tate two of the rolls back and forth under a lead weight (N). This may continue as much as two days, after which the needles are washed in a tub of hot soapy water (Fig. 2), dried in a rotary dryer (Fig. 3), and inspected for bent or broken ones. All that remains is to give the point a final sharpening (Fig. 7).

Plate 189   Making Needles III

Vol. I, Aiguillier-Bonnetier.

Plate 189   Making Needles III

These workers are making one of the many kinds of special needles, in this case knitting needles used in a stocking frame. These needles are hooked at one end, like a crochet needle, and contain an eye just above the hooked point. The wire is twisted through the nails of the device shown in Fig. 1 and cut in appropriate lengths. These are heated (at the right). A neck is flattened (6), an eye punched into it (Fig. 3), and the point filed sharp (Fig. 2). The completed needles are then polished (Fig. 4) and washed in a rotating drum (Fig. 5).

Plate 190   The Brazier I

Plate 190   The Brazier I

This spirited brazier's shop specialized in cooking utensils and large vessels of brass or copper. One hopes that the artist has crowded all this hammering and riveting into so small a space simply to show the variety of products, and that in actuality the artisans had more room. But they seem intent on their tasks, most of which are self-explanatory. It might, however, be well to call attention to the tub of Fig. 8, made by riveting brass sheets; to the operation of tinning a casserole (Fig. 4); and to the wheel of Fig. 3 which turns the lathe of Fig. 2 in a way not altogether apparent.

Plate 191   The Brazier II

Plate 191   The Brazier II

The guild of braziers also produced copper plates (see Plate 379) and musical instruments—orchestral brass and kettle drums. The instrument under construction is a hunting horn. Several finished examples hang upon pegs. The body of the horn is made from a piece of copper which is hammered around an iron mandrel or form fastened to the wall. A flaring mouth is soldered on to the tube at the forge (Fig. 2). Now the horn has the form of a long, thin trumpet, like that carried by some medieval herald, and it remains to curl it upon itself like a snail. This is accomplished without buckling the tube by the ingenious expedient of filling the cavity with molten lead (Fig. 3), which softens the copper so that it can easily be bent and gives it enough body to be coiled smoothly (Fig. 4).

Plate 192   The Lead Founder I

Plate 192   The Lead Founder I

Just as the iron founder bought pigs from the blast furnace, so the lead founder or "plumber" imported from England bars known in French as "salmons" of lead (Fig. 7).

Lead is one of the easiest metals to work because of its low melting point and high malleability. It has, however, little tensile strength, and is more suited for casting and rolling than for forging or drawing. This founder melts his lead on the furnace, not a terribly hot furnace (Fig. 4), and then pours lead sheet on a casting table (Fig. 1). Two men pour and a third controls the flow with his fire-rake. When cool, the sheet will simply be rolled up for storage or shipping (Fig. 11).

Plate 193 The Lead Founder II

Ever since the Roman Empire, one of the chief uses of lead has been for pipe (Fig. 7), which in this shop is made by casting. The pipe mold (Fig. 3) is equipped with a stoppered pouring hole every foot or so to assure even distribution of the metal. The pile of saumons *before the furnace (Fig. 13) may suggest from their shape why they were called that. Lead scrap (Fig. 11) could also be remelted and cast as pipe. Cross-sections and perspectives of the furnace appear below.*

Plate 194 The Lead Founder III

A second, rather more archaic method of making pipe consisted of rounding strips of sheet lead (Fig. 1) and soldering the seam (Fig. 2). Apparently finished pipe was gauged by weight (Fig. 3). The bottom of the plate illustrates the tools used in pouring sheet lead.

Plate 193   The Lead Founder II

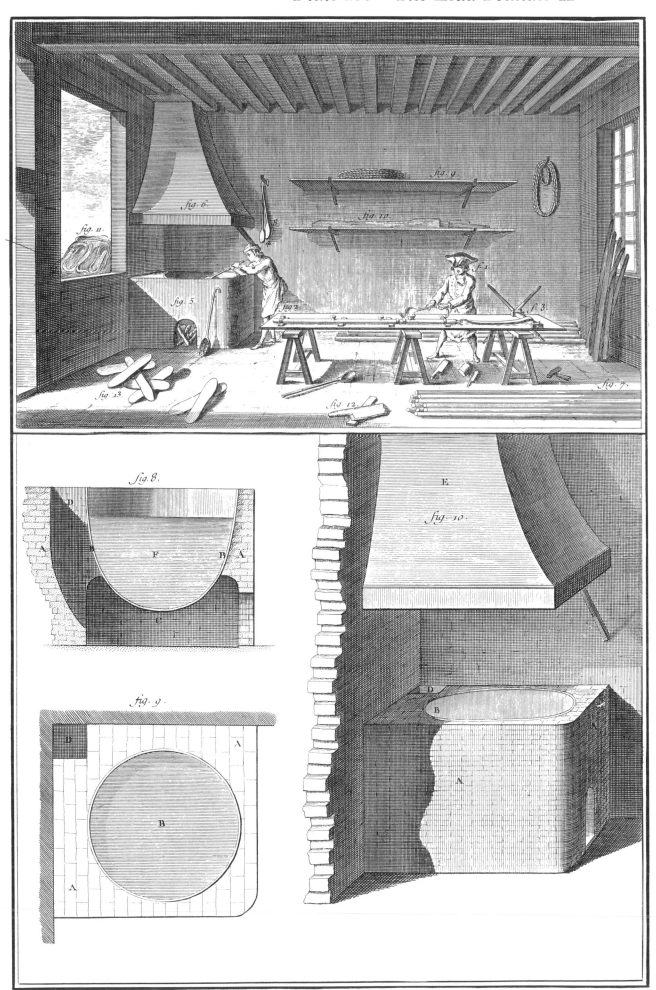

Plate 194   The Lead Founder III

Plate 195   The Lead Founder IV

To illustrate the rolling of lead, the Encyclopedists went to a much more mechanized foundry equipped to handle the metal on a large scale. Casting is completely mechanized—the cross-section in the lower right (Fig. 1) illustrates how the furnace is tapped directly into the pouring ladle. The ladle itself is tipped by a pair of beams. (There should be two men on the near beam but one of them has been "suppressed" like the dormouse in Alice because he would have obscured the view). The foreman has only to supervise and skim slag and foreign matter from the molten surface of the liquid metal. Everything has been thought of—even to the inclusion in the table of a stud (H) which will leave an eye in the solidified slab. Thus the tackle can be attached, and the crane used to slide the lead from the table. The block of lead must weigh tons.

Plate 196   The Lead Founder V

The crane (PSR) swings the slab casting to the rolling table (AB). There it is passed back and forth between rollers, which the foreman moves closer and closer together with every trip. Power is supplied by a water wheel outside the back wall of the shop, and the mechanism for reversing the direction of the rollers is illustrated in enlargement below (Fig. 6). When the sheet has been rolled thin enough to be used, say, for a lead roof, it is rolled up and again hoisted off the table by the crane.

Lead found other uses besides pipes and roofing. The growing chemical industry needed lead vessels. Pewter, the alloy of lead with tin, served as the poor man's silver. And the lovely soft gray surface of lead, which resisted the effects of weather and exposure better than any other metal, created a market in statuary, particularly garden statuary, and ornamental metal work.

Plate 195

Fig. 4.

Fig. 1.

12 Pieds

6 Pieds

Plate 196

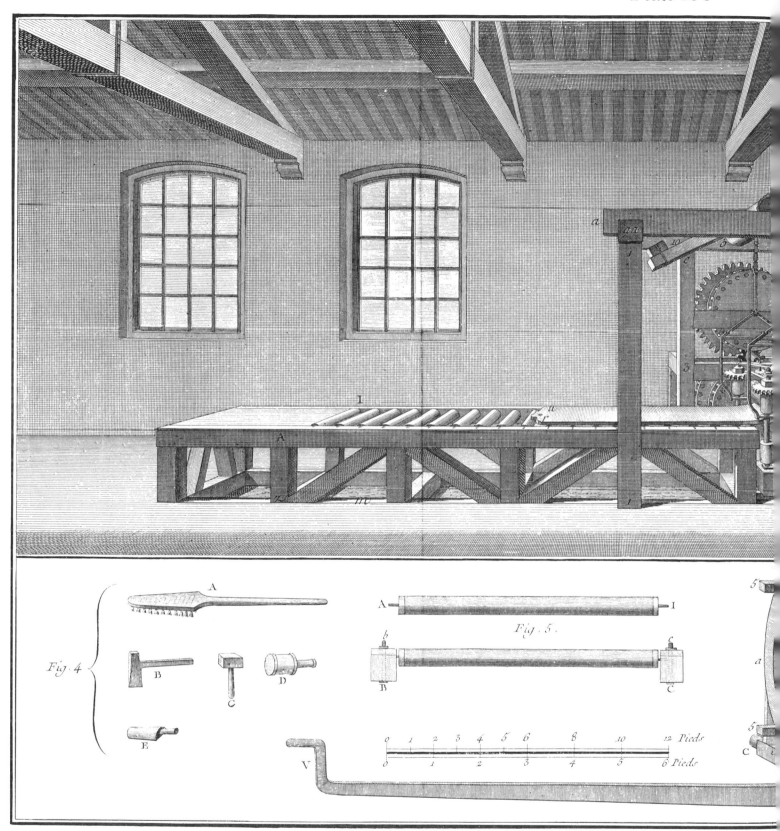

Fig. 4

A

B

C

D

E

Fig. 5.

A   I

b   c
B   C

V

| 0 | 1 | 2 | 3 | 4 | 5 | 6 | | 8 | | 10 | | 12 Pieds |

| 0 | 1 | 2 | 3 | 4 | 5 | 6 Pieds |

Plate 197

Plate 197   Pewter

Pewter vessels are highly admired nowadays, both for the unassuming restfulness of their surfaces and the modest simplicity of their forms. These were not qualities admired in the 18th century world of fashion, which demanded brilliance and embellishment, and the pewter-maker worked for humble customers. The writer, for example, once had the good fortune to be given a little pot au vin *(exactly like that of Fig. 1 facing) by a Parisian friend who explained that in older and more spacious days each of the servants in his family had been provided with such a utensil for the daily measure of wine.*

In the pewter-maker's workshop, two of his men turn a vessel on the lathe (a, b); another shapes a handle (c); the furnace is near the window, where a hot soldering iron is being withdrawn from the fire; and in the foreground the molder pours a ladle full of molten pewter.

Pewter

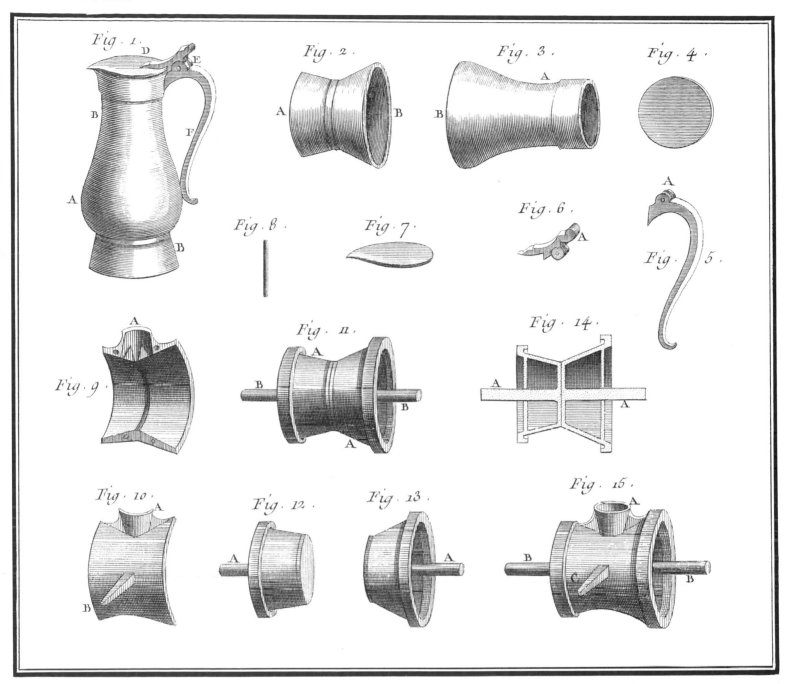

Plate 198   Coining I

A history of France has recently been published which sets off the rise of French civilization through the ages against the depreciation of French money. This process began with Charlemagne and was only temporarily interrupted in the 19th century, for the end is not yet, alas.

Just as modern French currency is more pleasing in appearance than in purchasing power, so in the 18th century the financial problem of the government was economic rather than metallurgical. It was not the fault of the mint that the state for which it produced fine coins was in chronic financial distress. The French livre *(pound) was down to one seventy-eighth of its original value in silver by the 18th century. All moneys have declined in purchasing power, of course, but even so the experience of the French* livre *contrasts strikingly with that of the English pound sterling, which was held comparatively firm across the centuries by the sobriety of the treasury in London.*

The machine for making money was a French invention, first in use in Paris in 1553, regularly employed from 1640, but adopted in London only in 1622. Its principle was that it rolled sheets from which coin blanks were stamped. It had been the medieval practise to hammer coins at a forge. This shop is designed for casting bullion from the furnace (c) in molds of sand (a).

Plate 198   Coining I

Plate 199   Coining II

Plate 199   Coining II

The most modern gold furnaces were blown by a bellows. Gold was melted in crucibles in small quantities. Both gold and silver coins were actually alloys containing small amounts of copper to make them hard enough to withstand use.

Plate 200   Coining III

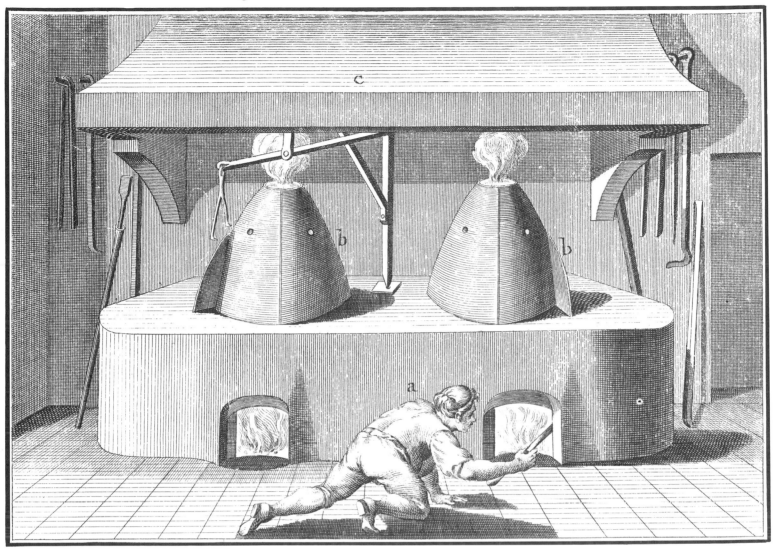

Plate 200   Coining III

Frenchmen hoard gold, but the word for money is argent *or silver, on which metal French money was based in practice. Silver bars were cast from a furnace like this, which was really a reverberatory hood on a wood stove. The crucible fitted under the hood.*

Plate 201

Plate 201   Coining IV

The power source of this rolling mill is illustrated more clearly than the rollers them-
selves, which can be seen in the top-view of the machine given on the opposite page.
Ingots were fed vertically between rollers at the left (A), and then horizontally between
two more narrowly spaced pairs (B). This process rolled bars of precious metal into
long, thin strips from which coin blanks were stamped.

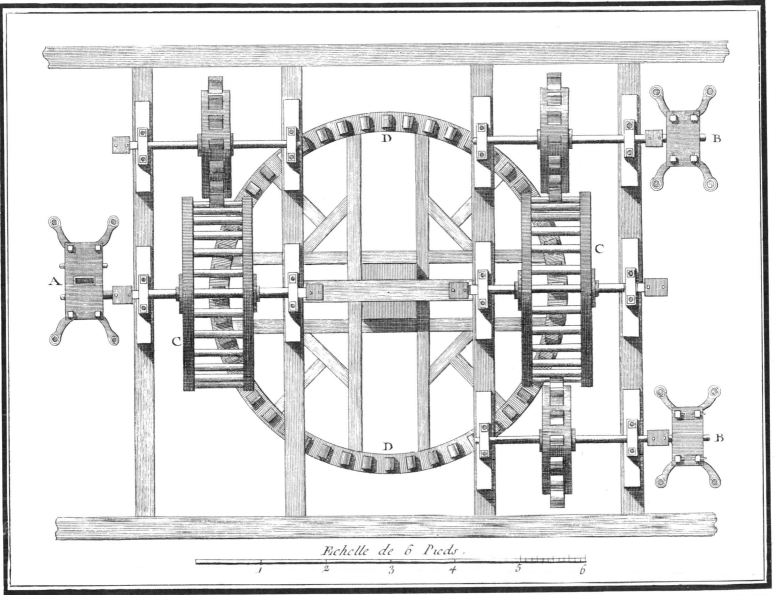

Echelle de 6 Pieds.

1   2   3   4   5   6

Plate 202   Coining V

Plate 202 offers a perspective view of the two types of roller (Figs. 1 & 13). They were not large. Each roller was only about six inches in diameter. But that they acted powerfully will be apparent if it is remembered that all the strength of four spirited horses, multiplied by the mechanical advantage of the gear ratio shown in the previous plate, was exerted to turn these three small mills which rolled heated ingots into strips of precious metal the thickness of a coin.

Plate 203   Coining VI

Plate 203, Fig. Ier is the machine for punching out blanks. The artist appears to have made a mistake in representing the basket (L) as full of figured coins—at this stage they should be only discs of gold or silver.

Plate 202 Coining V

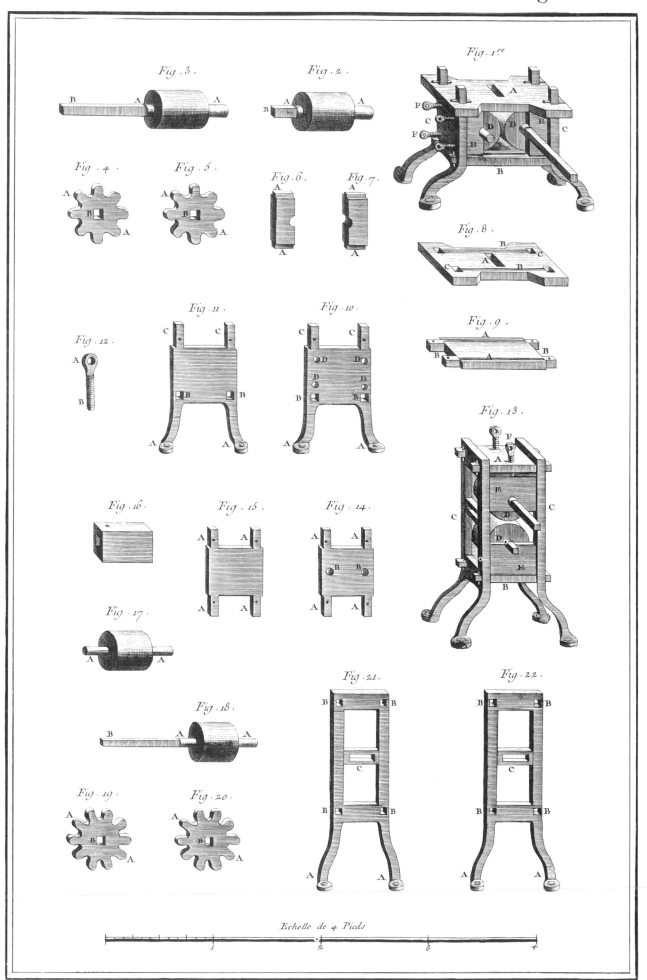

Plate 203   Coining VI

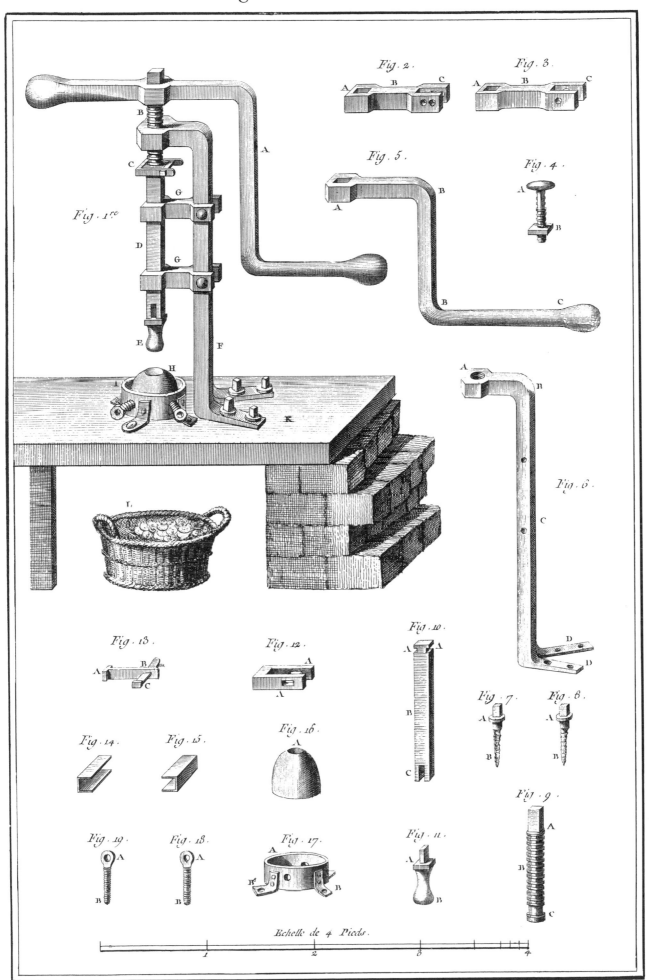

Fig. 2. Fig. 3.

Fig. 5. Fig. 4.

Fig. 1re

Fig. 6.

Fig. 13. Fig. 12. Fig. 10.

Fig. 14. Fig. 15. Fig. 16. Fig. 7. Fig. 8.

Fig. 9.

Fig. 19. Fig. 18. Fig. 17. Fig. 11.

Echelle de 4 Pieds.

Plate 204

Plate 204 Coining VII

Before they could be struck into coins, blanks had to be annealed in the reverberatory annealing furnace. The upper picture is a side-view and (opposite) a section through the middle of the furnace.

Plate 205   Coining VIII

Plate 205   Coining VIII

The press is fitted with two matrices. One prints the head of Louis XV on one side of the coin, and the other the "escutcheon," corresponding to our "tails." Blanks from basket (P) are stamped one by one and tossed into the second basket (Q).

Plate 206   Coining IX

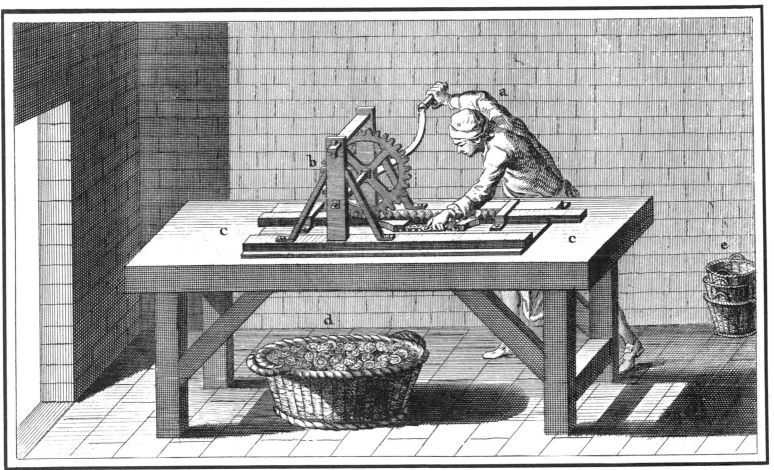

Plate 206   Coining IX

Milling the circumference of the coin prevents clipping, the practice of medieval money changers who would snip a little gold from the edge of a crown piece and pass the remainder for full value.

Plate 207

Plate 207 Coining X

Finally silver coins were blanched and very thoroughly "pickled" in the French mint, which used a weak tartaric acid to make its coins all bright and shining before they started down in the world.

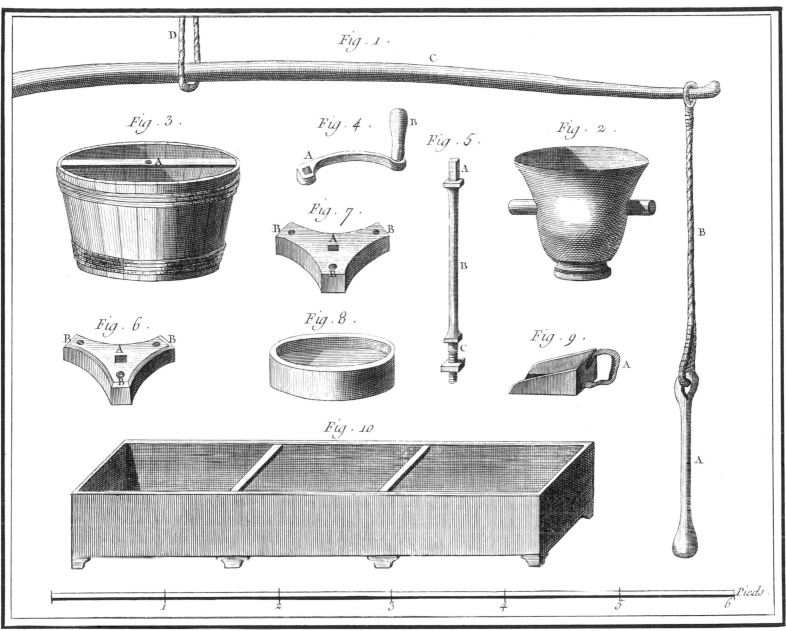

Fig. 1.

Fig. 3.

Fig. 4.

Fig. 5.

Fig. 2.

Fig. 7.

Fig. 6.

Fig. 8.

Fig. 9.

Fig. 10.

Pieds

1   2   3   4   5   6

Plate 208

Plate 208   Coining XI

A table of some of the principle 18th century coins:

1 & 2	The silver écu of France.
3 & 4	The "Louis d'or—Golden Louis" of France.
5 & 6	The silver piastre of Spain.
7 & 8	The golden pistole of Spain.
9 & 10	The silver cross of Portugal.
11 & 12	The gold piece of Portugal.
13 & 14	The silver crown of England.
15 & 16	The golden guinea of England.
17 & 18	The rix-dollar of Holland.
19 & 20	The golden ruyder of Holland.
21 & 22	The golden ducat of the Austrian Netherlands (Belgium).
23 & 24	The golden sovereign of the Austrian Netherlands.

Fig. 9. Fig. 10. Fig. 11. Fig. 12.
Fig. 13. Fig. 14. Fig. 15. Fig. 16.
Fig. 17. Fig. 18. Fig. 19. Fig. 20.
Fig. 21. Fig. 22. Fig. 23. Fig. 24.